D1568933

Trees and Networks
in
Biological Models

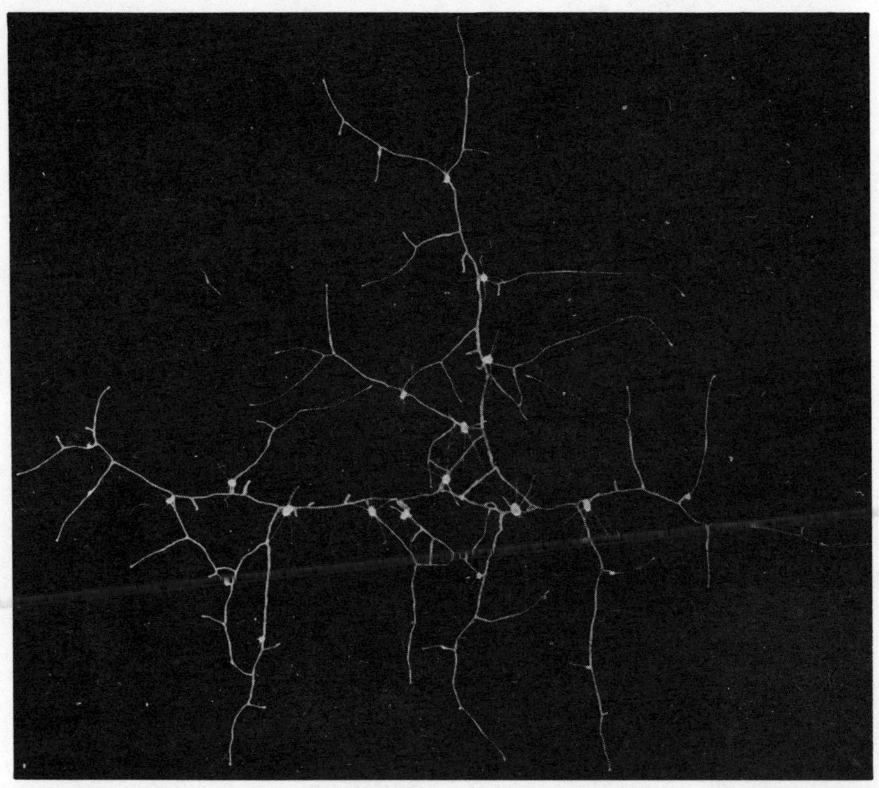

Frontispiece. A 9-day old colony of the hydroid *Podocoryne carnea*, grown on a microscope slide in sea water. Photograph kindly supplied by Dr. Braverman and published with his permission. © 1978 by M. H. Braverman

Trees and Networks in Biological Models

N. MacDonald

Department of Natural Philosophy
University of Glasgow, Scotland

A Wiley–Interscience Publication

JOHN WILEY & SONS

Chichester · New York · Brisbane · Toronto · Singapore

Library of Congress Cataloging in Publication Data:

MacDonald, N. (Norman), 1934–
Trees and networks in biological models.

"A Wiley-Interscience publication."
Includes index.
1. Biological models. I. Title.
QH324.8.M2 1984 574′.0724 83-7018
ISBN 0 471 10508 2

British Library Cataloguing in Publication Data:

MacDonald, N.
Trees and networks in biological models.
1. Biological models
I. Title
574′.0724 QH324.8

ISBN 0 471 10508 2

Typeset in Great Britain by
Pintail Studios Ltd, Ringwood, Hampshire
Printed by The Pitman Press, Bath, Avon

Contents

Preface ix

1. Introduction: Real and Abstract Trees and Networks 1
1.1 Static and dynamic treatments 2
1.2 Terminology of graph theory 4
1.3 Summary of this volume 6

PART I DOMINANCE, PREDATION, AND COMPLEXITY

2. Hierarchy in the Relationships of Individuals and Species 11
2.1 An index of hierarchy 12
2.2 A model of intrinsic traits 15
2.3 The dynamic interaction digraph 16

3. The Trophic Structure of Foodwebs 20
3.1 Trophic impurity 21
3.2 Reduction of a weighted foodweb to a food chain 23

4. The Interval Structure of Foodwebs 31
4.1 Continuous and discrete resources 31
4.2 The consumer graph and the resource graph 32
4.3 The interval property 35
4.4 Interpretations of the interval property 38

5. Complexity of Trees and Networks 40
5.1 Trees 42
5.2 Networks 42

PART II STABILITY OF EQUILIBRIUM STATES AND OF PERIODIC BEHAVIOUR

6. Stability and Complexity of Model Foodwebs 49
6.1 Two types of model 50
6.2 Stability and connectedness 52
6.3 Self-stability, omnivory, compartmentalization 53

7. Graphical Aspects of Local Stability Theory: Cycle Analysis 56
7.1 The secular equation 58
7.2 Difference equations 62

8. Applications of Cycle Analysis 66
8.1 Sign stability 66
8.2 Competition and predation 67
8.3 A measure of relative stability 68
8.4 Distributed delays 69
8.5 The Goodwin oscillator 72
8.6 A seedbank problem, and some observations on rosettes 74

9. Power Laws and Switching Functions 78
9.1 Power laws 78
9.2 Switching functions 80
9.3 Stable cycles 83

10. Large Scale Clocked Switching Networks 89
10.1 Switching networks of Walker and Ashby 92
10.2 Kauffman's networks and the forcing property 94

PART III BRANCHING STRUCTURES: DESCRIPTION, BIOPHYSICS, AND SIMULATION

11. Branching Structures in Biology: Topology and Geometry 103
11.1 Dendritic trees 104
11.2 Lung airways 106
11.3 Arteries 108

12. Law and Order in Trees 110
12.1 Ordering methods 110
12.2 Horton's law of the branching ratio 115
12.3 Random binary trees 117

13. Branching Ratios, Branch Length Ratios, and Branch Diameter Ratios with Strahler Ordering 122
13.1 The ratios R, R_d, and R_l 122
13.2 Power laws 131
13.3 Some details of branching 135

14. The Lung as a Space-filling Tree 140
14.1 Symmetrical branching at 90° 140
14.2 Symmetrical branching at any angle 143
14.3 Preliminary ideas on asymmetry 146
14.4 Filling a surface with leaves 149

15. Branching Diameter Ratios and Branching Angles: Biophysics 151
15.1 The ratio R_d 151
15.2 Branching angles 155

16. Pipes, Bundles, and Horns: More Biophysics 160
16.1 The pipe model for a botanical tree 160
16.2 Rall's model of dendrites 163
16.3 Models for impedance calculations with arteries and airways . 165

17. Simulating the Growth of Dendritic Trees 168
17.1 Terminal and segmental growth models 169
17.2 An identity for terminal growth trees 173

18. The Mathematics of Tree Simulation: L-Systems 175
18.1 A simple tree and its string sequence 175
18.2 Context-free and context-sensitive L-systems 177
18.3 Recurrence rules and growth functions 180

19. Applications of L-Systems 184
19.1 A filamentous alga and a branching alga 184
19.2 Recurrence in Shreve orders 186
19.3 Inflorescences 189

20. Simulation of Growth with Anastomosis: A Colonial Hydroid 194
20.1 Consequences of anastomosis 194
20.2 Simulations of *Podocoryne carnea* 195
20.3 Some possible extensions 196

Appendix 1 Simplicial complexes 200

Appendix 2 Linear homogeneous difference equations 202

Appendix 3 Laminar flow in a tube 204

Appendix 4 Details of the analysis of branching angles 206

Appendix 5 Impedance and the reflection of waves 209

Subject Index 212

Preface

This book arose from my interest in certain kinds of abstract network in biological models, such as foodwebs in ecology and linked control loops in biochemistry. This led me to consider a review emphasizing the evaluation of numerical indices to characterize features of networks, and the effects of closed loops of interactions on stability of equilibria and of periodicity. This developed into Parts I and II. On reading more about the application of graphs in biology, it became apparent that there is a particular need for a review of models and simulations of real trees and networks. Part III provides such a review.

I hope that this book will interest graduate students and research workers in biology who have an interest in some kind of real or abstract tree or network, and who feel the need for a comparative study of these structures. I hope also it will be read by mathematicians looking for suitable areas of application in biology.

The mathematical level is intended to be not too high for biologists with an interest in modelling. The language and some of the basic concepts of graph theory are used throughout. The most technical sections, confined to Part II, are concerned with local stability of equilibrium.

Some of the material for this book was used in lectures and informal talks in Glasgow and Dublin. I wish to thank various colleagues for comments: Jagan Gomatam, Hedley Morris, Roger Nisbet, Neil Spurway, John Usher and George Wyllie. Roger Nisbet, in particular, made many suggestions for the improvement of Parts I and II. Thanks are due to George Sugihara and R. E. Ulanowicz for sending unpublished material, and to M. H. Braverman for sending a number of his photographs of colonial animals. Others who have permitted reproduction of tables and figures are acknowledged separately.

I wish to thank Margaret MacDonald for some drawings, and for improving the clarity of the text at many points, Ian MacVicar for photographs, Ewan Kirk for a computer program to draw trees, and Kay Johnston for typing.

N. MacDonald
January 1983

Chapter 1

Introduction: real and abstract trees and networks

The habit of analysing complex systems into discrete components and their interactions is deeply ingrained, and leads naturally to portrayal of the components as points and their interactions as lines joining pairs of points. It is natural also to think of a real branching structure in terms of a skeleton description in which branches are replaced by lines and their regions of bifurcation by points where two lines meet.

There is a basic distinction between two kinds of pattern built up of lines and points. In a network there are closed paths, in a tree there are none. Trees and networks abound in biology. There are many three-dimensional real branching structures, such as the dendrites of a nerve cell (Fig. 1.1), lung airways, and the branches of a botanical tree. (Since mathematicians have pre-empted the use of

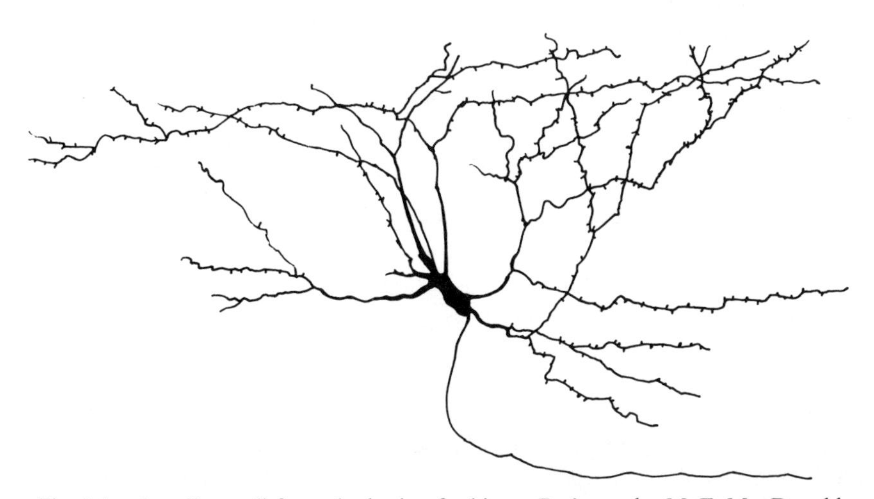

Fig. 1.1 A stellate cell from the brain of a kitten. Redrawn by M. F. MacDonald from an illustration in Ramon y Cajal, *Histologie du Système Nerveux*, Paris, 1909

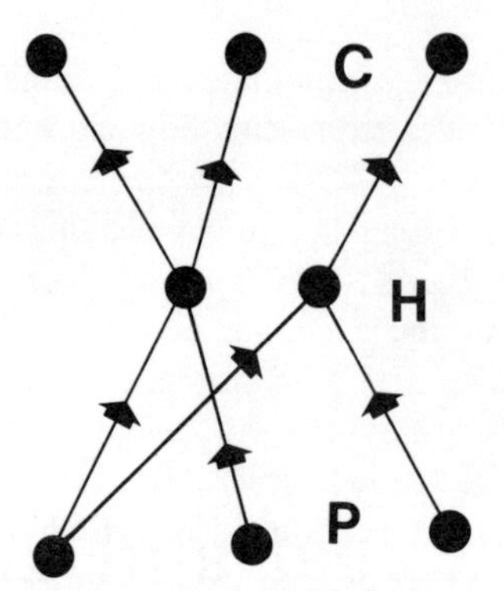

Fig. 1.2 A simple food web with three distinct trophic levels, carnivores **C**, herbivores **H** and plants **P**

'tree' as a technical term implying the absence of closed paths, I shall regretfully continue to use this clumsy locution for the kind of tree you can climb.) In some cases, such as arteries and certain colonial animals, the branching structure becomes a network by anastomosis (*Frontispiece*). Abstract trees are employed in taxonomy, and abstract networks in a variety of biological contexts. For example, we can enumerate the pairwise relations '*i* eats *j*' between species sharing a common habitat, thus describing a foodweb. Similarly we can enumerate cases of competition between species.

Predation is a directed relationship while competition is a symmetric one. Because of this distinction we may wish to give the lines a preferred direction. For example, in Fig. 1.2 we express the relationships between carnivores C, herbivores H, and plants P by lines on which the arrows follow the direction of energy transfer through this simple web. Again, in studying the dominance relations in an animal society, such as the classic case of a flock of domestic fowl, we can use lines directed from the superior to the inferior of each pair.

1.1 STATIC AND DYNAMIC TREATMENTS

Both real and abstract trees and networks can be described as static structures. A descriptive terminology, or a set of numerical indices, may be sought to characterize particular aspects of this static structure. For example, we may describe a certain botanical tree as having six orders of branches—from the first, comprising twigs carrying leaves, to the sixth, consisting solely of the main trunk. We may also count the number of branches belonging to each order and examine whether these data can be summarized in terms of a single branching ratio, between the number in one order and the number in the next higher order. We may describe a flock of domestic fowl as exhibiting a hierarchical structure, or strict pecking order, in which one member dominates all the rest, a second dominates all but the first, and so on. If the relations are not completely ordered in this way, so that, for example, 1 dominates 2, 2 dominates 3 and 3 dominates 1, we may seek a numerical index expressing how close the structure comes to strict hierarchy.

We would soon find it necessary to go beyond this static description. We might wish to examine the growth of a tree from a single shoot to a many-branched structure. Also, by adopting the schematic description in terms of points and lines, in which the relative lengths and orientations of lines are ignored, as well as the thicknesses of real branches, we throw away most of what distinguishes one kind of real branching structure from another. The examples cited have in common repeated division (usually bifurcation, although in botanical trees more complicated divisions occur) into finer, and usually shorter, branches. But their functions and environments differ greatly. Lung airways or arteries can reasonably be treated as isolated branching structures. (Even here we may ask how the veins echo the arteries or how the pulmonary arteries echo the airways.) But nerve cells reach out their dendrites to make contact with the axons of other nerve cells, and typically the terminal segments of dendrites are longer than those nearer the nerve body. The branching pattern of an isolated botanical tree is affected by the need to cover an area with leaves uniformly in order to use sunlight efficiently. The branching pattern of the lung airways is influenced by the need to locate alveoli (the sites of gas transfer) uniformly throughout the lung volume. These requirements, as we shall see, lead to branches that are shorter as one goes outwards from the trunk or the trachea through successive orders.

The botanical tree has to support its own weight, and the lengths and thicknesses of the branches are influenced by this constraint. This is less important for the lung, although the blood vessels in the lung have a secondary mechanical supporting role. This constraint is presumably not a major consideration for nerve cells. Lung airways, arteries and veins, and the branches of botanical trees transfer fluids, by pumping or osmosis. Dendrites carry electrical signals. As we move from a topological description, in which all we know is the configuration of points and lines, to a geometrical description in which lengths and diameters of branches, and angles of branching, are specified, we need to consider these mechanical and functional requirements.

With abstract networks we frequently have to replace a static by a dynamic description. In the foodweb, even if we regard all species as permanent, their populations will fluctuate, and we may wish to model population changes. If we use a model constructed in terms of ordinary differential equations, for example, this will contain functions expressing the effect on the rate of change of a prey population of the presence of the predator population, and the reciprocal effect on the rate of change of the predator population. We can still summarize the effects by a diagram analogous to Fig. 1.2, but now there are lines directed each way between any pair of points and time-varying numerical values associated with them. The interactions may not be permanent; for example, a predator may lose interest in one of its prey species if the population of that species falls below a certain threshold.

The dominance relations in a flock may change with time, and the static structure can only be a snapshot of a particular phase in a changing set of relationships. This leads to another kind of description in terms of points and lines. We can define a state of the flock (with reference only to dominance) by listing all the directed interactions $1 \to 2$, $1 \to 3, \ldots, 2 \to 2, \ldots$. For N members there are $\frac{1}{2}N(N-1)$ such interactions, each of which can be in a '+' state (such as $1 \to 2$) or a '−' state (such as $2 \to 1$). So there are a finite number of states, in fact

$$2^{\frac{1}{2}N(N-1)}$$

of the whole flock. Let us denote each such state by a point, and a transition from one state to another by a line directed from the first point to the second. Since transitions must correspond to a single reversal of dominance, only a limited number of lines are possible. A sequence of reversals of dominance gives a path through this diagram. The actual path followed by the state will have a random element, since at each encounter there will be a certain probability of a dominance reversal. This approach to the dynamics of a system of interacting components can be extended to cases in which at first sight the number of states is not finite. The trick is to regard a transition from low values to high values of a continuously variable quantity as if it were an abrupt switch from the state 'low' to the state 'high'.

1.2 TERMINOLOGY OF GRAPH THEORY

The natural mathematical language for this style of abstract description of the structure or dynamics of a system is that of graph theory. It is convenient to gather together here a number of definitions of terms in graph theory used repeatedly in this book. More specialized terms will be introduced as needed. The terminology of graph theory has not entirely crystallized and some arbitrary choices have to be made. A **graph** G consists of a finite set $X = \{X_1, X_2, \ldots, X_V\}$ of V members called **vertices**, and a set of pairs (X_i, X_j) called **edges**. The graph can be represented by a diagram in which a vertex X_i is a point labelled i and an edge (X_i, X_j) is a line ij connecting points i and j. The graph can also be represented by a $V \times V$ matrix with entries $A_{ij} = A_{ji}$ which are 1 if there is an edge ij and 0 otherwise. This is called the **adjacency matrix**, because i and j are said to be **adjacent** if there is an edge ij.

While the pictorial version of the graph has manifest attractions in revealing aspects of the structure of graphs with low V, the matrix version allows explicit algebraic results to replace the tedious process of counting paths through a graph when V is large.

A directed graph or **digraph** D consists of a finite set $X = \{X_1, X_2, \ldots, X_N\}$ of N members called **nodes**, and a set of ordered pairs (X_i, X_j) called **arcs**. The digraph can be represented by a diagram in which each X_i is represented by a point labelled i, and each arc (X_i, X_j) by a line (i, j) directed from i to j. Both (i, j) and (j, i) may appear. Points i and j are said to be **adjacent** if either (i, j) or (j, i) is present. Again, an **adjacency matrix** can be defined, in which $A_{ij} = 1$ if (i, j) is present, $A_{ij} = 0$ otherwise. It is sometimes useful to assign to each arc a numerical value (**weighted digraph**) or a sign $\pm$ (**signed digraph**).

Graphs and digraphs are sometimes defined to exclude the edge *ii* or arc (i, i). However, I shall allow such **loops**, and the corresponding diagonal elements A_{ii} of the adjacency matrix.

I shall use the term **contiguous** for an edge *ij* and the vertex *i* or *j*, or for two edges *ij* and *jk*. Similarly I shall use this term for an arc (i, j) and node *i* or *j*, or for two arcs (i, j) and (j, k) or (i, j) and (k, j). (Two arcs can be contiguous even if they are directed with the arrows against one another.) A **path** is a set of contiguous arcs or edges. A path revisiting a node or vertex is a **cycle**. For example, *ij*, *jk*, *ki* is a cycle on a graph, and (i, j) (j, k) (i, k) (in which the arrows are *not* all aligned) is a cycle on a digraph. The **length** of the cycle is the number of edges or arcs it contains. A **loop** is a cycle of length 1.

A **tree** is a graph with no cycles. A **rooted tree** is a tree in which one vertex is distinguished as the **root**. Since practically every tree mentioned in this book is a rooted tree, I shall usually drop the adjective. A **leaf** is a vertex of a rooted tree that is contiguous to only one edge and is *not* the root. A **binary tree** is a tree in which at most three edges are contiguous to any vertex. Fig. 1.3 illustrates a number of these terms.

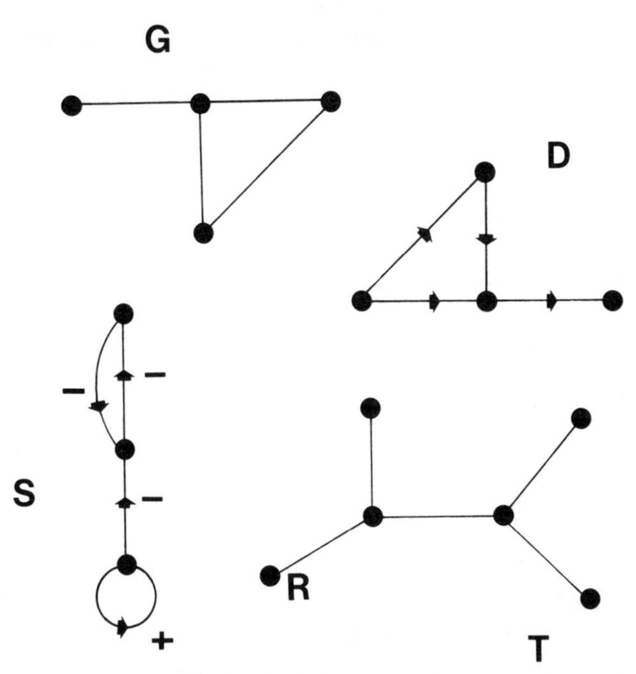

Fig. 1.3 Some of the terminology of graph theory. **G** is a graph with 4 vertices and 4 edges. **D** is a digraph with 4 nodes and 4 arcs. **S** is a signed digraph with 3 nodes and 4 arcs, one of which is a loop. **G** and **D** each contain a cycle of length 3, **S** contains a cycle of length 2. **T** is a rooted tree, with root **R** and five leaves

Graphs are used when the relationship is symmetrical. Digraphs are used when relationship is one-way, or when it is two-way with some distinction between the (i, j) and (j, i) interactions. Digraphs are also used when describing state transitions, since as time elapses the system goes from a specific state to another specific state. Rooted trees are appropriate as a shorthand description of biological branching structures without anastomosis, since we almost always have an identifiable starting point, such as trunk, trachea or nerve body.

1.3 SUMMARY OF THIS VOLUME

This book consists of three parts, each of which is fairly self-contained. The first deals with abstract networks, in particular those concerned with dominance and predation. The second deals with dynamics, first with the local stability of equilibrium points in models of interacting components, and then with the stability of oscillations. The third part deals with the description, biophysics and simulation of real branching structures. I should like to emphasize that this part, which is the longest, is essentially independent of the first two.

Chapter 2 deals with dominance, which is a significant concept both in connection with social interactions of individual animals and in connection with the competition of species for a shared resource such as living space. This topic is introduced first because it serves to illustrate both the static description of a set of relationships as a digraph, and the dynamic description of successive changes in the relationships as a path on a digraph. An index of hierarchy is introduced.

Chapter 3 deals with the trophic structure of foodwebs, that is their approximate division into subsets such as producers, herbivores, and carnivores. An index of trophic impurity is introduced. A process is described for reducing a foodweb, in which energy transfers are quantified, to an approximate equivalent food chain.

Chapter 4 is concerned with a more abstract property of a foodweb, the interval property. This is defined in the context of treating a shared prey as evidence of competition. Different interpretations of this property are discussed.

Chapter 5 discusses measures of complexity for rooted trees and graphs with many cycles.

Chapter 6 begins the second part of the book and also links it with the first part. It examines models of foodweb dynamics formulated in terms of sets of coupled ordinary differential equations (o.d.e.). The local stability of an equilibrium point in this kind of model is discussed in relation to the connectivity (a particular measure of complexity) of the web and in relation to its trophic purity.

Chapter 7 continues the study of local equilibrium of models using o.d.e. or difference equations. When the dynamics of a system is described in a linearized version, appropriate near an equilibrium point, there is an unambiguous weighted digraph structure associated with the interactions. While local stability theory might seem to involve merely the operation of mechanical algorithms, such as the Routh–Hurwitz tests for o.d.e., some insight can be gained by studying the cycle structure of this digraph. Chapter 8 considers applications of this cycle analysis to models relating to annual plant populations, predation and biochemical control loops.

Chapter 9 examines a way of qualitatively understanding the periodic behaviour of solutions of equations modelling coupled control loops. The continuous activation and inhibition functions are replaced by Boolean functions (on–off switches). This reduces the state space to a finite set of points, and the development of the system in time to a path along the arcs of a digraph. Since the set of points is finite and the dynamics deterministic, only steady states and periodic solutions can be treated.

Chapter 10 discusses how to extend this Boolean approach so that it applies to very large systems. The trick here is to synchronize all the switches. The question particularly examined is how the dynamics of a system with N components and 2^N states can be dominated by short cycles (period $\sim NT$, where T is the time step) rather than long ones (period $\sim 2^N T$).

In the third and longest part of the book, Chapters 11–14 are mainly concerned with the static description of rooted trees, while Chapters 15–20 discuss dynamics of flows and signals in trees, and the growth of trees and (anastomosing) networks.

Chapter 11 introduces real branching structures, illustrating how it is essential to isolate topological aspects (branch ordering and branch counting) from geometrical aspects (such as branch length and branching angle).

Chapter 12 discusses ordering methods introduced by geographers and by physiologists and includes a discussion of randomly constructed rooted binary trees. The work of geographers on stream networks between 1950 and 1970 had a considerable influence on subsequent network analysis by biologists.

Chapter 13 presents a considerable amount of data on branch numbers, diameters and lengths, employing only the most popular ordering system, that of Strahler, so that comparisons can be made between different types of branching structure. Some power laws linking the ratio of branch numbers in successive orders, R, and the ratios of mean branch lengths or diameters in successive orders, R_l and R_d, are introduced. These are suggested by models discussed in the subsequent chapters.

Chapter 14 discusses the lung as a branching structure uniformly filling a volume with branch tips. This leads to a tentative relationship between R and R_l. There is also a brief discussion of the analogous property of botanical trees.

Chapter 15 considers the biophysics of the flow of blood in arteries and gas in airways, which leads to predictions about the relationship of R and R_d, and about branching angles.

Chapter 16 discusses 'pipe' or 'bundle' models in botany, in electrophysiology of dendrites, and in wave propagation in arteries or lung airways. In these models branching angles are ignored and the tree is replaced by a single tube or a bundle of parallel, although bifurcating, tubes. These models again suggest relationships between R and R_d.

Chapter 17 discusses the simulation of the growth of dendritic trees using local hypotheses about the initiation of branches, and testing these hypotheses against data on the peripheral parts of a mature dendritic tree.

Chapter 18 introduces a formalism for tree simulation, the L-system method. Again the emphasis is on testing local growth hypotheses against global topology. Chapter 19 describes applications of L-systems to the simulation of branching algae and inflorescences.

Chapter 20 discusses questions raised by the simulation of a network with frequent anastomoses, the stolon network of a colonial hydroid.

Part I

Dominance, Predation, and Complexity

'At the head of everything is God, Lord of Heaven. After Him comes Prince Torlonia, lord of the earth. Then come Prince Torlonia's armed guards. Then come Prince Torlonia's armed guards' dogs. Then nothing at all. Then nothing at all. Then come the peasants.'

I. Silone, *Fontamara*

'All things have their uses and their part and proper place in Nature's economy: the ducks eat the flies—the flies eat the worms—the Indians eat all three—the wild cats eat the Indians—the white folks eat the wild cats—and thus all things are lovely.'

M. Twain, *Roughing it*

Chapter 2

Hierarchy in the relationships of individuals and species

In an early classic of ethology T. Schjelderup-Ebbe (1922) described the dominance relations between domestic fowl, expressed by priority in feeding, or pecking order. When any two hens of a flock meet, one will always give way to the other. In flocks of ten to twenty hens it is possible to enumerate the dominance relations completely, and to demonstrate that they are strongly hierarchical. Cyclic relationships (i dominates j, j dominates k, k dominates i) are relatively rare and most of the hens can be ordered in terms of the number of their companions which they dominate. An example cited by Schjelderup-Ebbe is of a flock of ten hens, three of which dominated eight others, one dominated six, one five, and so on. In this flock the three leading hens had a cyclic pattern of dominance but otherwise there was a strict hierarchic order. Similar dominance patterns have been recorded in species as diverse as ants (Cole 1981) and macaque monkeys (Schulman and Chapais 1980).

Hierarchic structure within a population can have significant effects on overall population stability (Lomnicki 1978) and on the effectiveness of a population as a competitor with or predator on other species (Gurney and Nisbet 1979). These authors assign to each member of a population a rank and assume, for example, that the reproduction rate depends on rank. So they assume a strictly hierarchic ordering. It is of interest to quantify how closely real populations approach this.

Dominance relationships can be significant between populations as well as within them. Buss and Jackson (1979) present data on dominance relations observed in competition between twenty species of marine organism occupying the underfaces of corals in a reef on the north coast of Jamaica. Here dominance is expressed by one species overgrowing the other. For some pairs a clear dominance relationship is not established. In different samples a can overgrow b or b overgrow a. Excluding these pairs, there are still subsets of the species within which the relations are quite far from strict hierarchy. One of these subsets consists of an ascidian, five sponges and a coralline alga, among which two species overgrow five others, two overgrow three, two overgrow two and one overgrows one other.

2.1 AN INDEX OF HIERARCHY

As a preliminary to attempts to model the process of setting up or maintaining a dominance pattern, Landau (1951a,b) described how to obtain a single numerical index to express the extent to which the pattern differs from strict hierarchy. Even in a static description (Landau 1951a) it is obvious that much information is discarded in reaching a single index. In a dynamic description (Landau 1951b) of fluctuating dominance relations this discarded information can be essential. In relation to the contents of this volume the importance of Landau's work is in its illustration of three general features:

(a) the use of a numerical index to quantify a global property of a system, at the cost of discarding possibly valuable detail;
(*b*) the use of this index in testing a static model more economically than if one had to look at all the relationships in detail;
(*c*) the use of a dynamic model employing a digraph.

Let us first consider the static digraph as used in Landau's first paper. The flock is associated with a set of N labelled nodes i and the dominance relations by the arcs (i,j), where i dominates j. There is always a definite outcome of any encounter, so the digraph is fully connected. That is to say, there is either an arc (i,j) or an arc (j, i) for each pair of nodes i and j. The preponderance of hierarchic patterns means that, for example, a digraph with four nodes is more likely to resemble **A** than **B** in Fig. 2.1. In describing fixed dominance patterns the labels

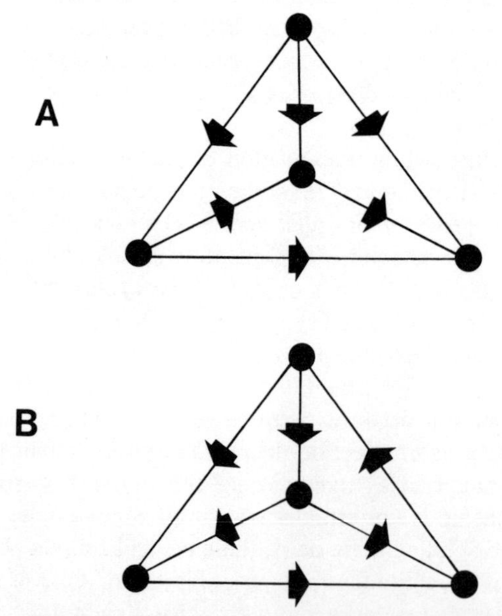

Fig. 2.1 Digraphs exhibiting hierarchic (**A**) and democratic (**B**) patterns of dominance

for the nodes can usually be dropped. While the pictorial aspect of the digraph can be useful it is not essential, and one can work entirely in terms of the adjacency matrix A. For a digraph with N nodes this is an $N \times N$ matrix with elements

$$\begin{aligned} A_{ii} &= 0 \\ A_{ij} &= 1, \text{ if there is an arc } (i,j) \\ &= 0, \text{ otherwise} \end{aligned}$$

So the digraphs **A** and **B** of Fig 2.1 have respectively the adjacency matrices

$$A_1 = \begin{bmatrix} 0 & 1 & 1 & 1 \\ 0 & 0 & 1 & 1 \\ 0 & 0 & 0 & 1 \\ 0 & 0 & 0 & 0 \end{bmatrix}, \quad A_2 = \begin{bmatrix} 0 & 1 & 1 & 0 \\ 0 & 0 & 0 & 1 \\ 0 & 1 & 0 & 1 \\ 1 & 0 & 0 & 0 \end{bmatrix}$$

The sequence of rows is immaterial, but the hierarchic nature of the diagraph **A** comes out most clearly by presenting A_1 in the form which has an upper right triangle of 1s.

The process of boiling down the information contained in the adjacency matrix into a single index goes in two steps. In the first a vector of dimension N is derived from the matrix, in the second a single number is derived from the vector. Information is lost at each step. The **score sequence vector** $\boldsymbol{V}$ has components which are the row sums of A, taken in non-increasing order. These components are the **out-degrees** of the nodes of the digraph, that is to say, the numbers of arcs leading away from each node. Thus for the digraphs **A** and **B** of Fig. 2.1, with adjacency matrices A_1 and A_2, the score sequence vectors are respectively

$$\boldsymbol{V}_1 = [3, 2, 1, 0], \quad \boldsymbol{V}_2 = [2, 2, 1, 1]$$

If each member of a flock meets each of the others there are $\frac{1}{2}N(N-1)$ victories. So the fully connected digraph has $\frac{1}{2}N(N-1)$ arcs. The average value of a component V_i of the score sequence vector $\boldsymbol{V}$ is therefore

$$\langle V_i \rangle = \tfrac{1}{2}(N-1)$$

When N is odd one can construct a fully democratic digraph in which each V_i is equal to this average value. When N is even, the most democratic digraph possible has half the V_i equal to $\frac{1}{2}N$ and the other half equal to $\frac{1}{2}(N-2)$. The fully hierarchic digraph always has the score sequence vector

$$\boldsymbol{V} = [N-1, N-2 \ldots 2, 1, 0]$$

Information is lost when A is replaced by $\boldsymbol{V}$. For example the score sequence vector

$$V = [3, 2, 2, 2, 1]$$

could correspond either to A_3 or to A_4, where these adjacency matrices are

$$A_3 = \begin{bmatrix} 0 & 1 & 1 & 1 & 0 \\ 0 & 0 & 0 & 1 & 1 \\ 0 & 1 & 0 & 0 & 1 \\ 0 & 0 & 1 & 0 & 1 \\ 1 & 0 & 0 & 0 & 0 \end{bmatrix}, \quad A_4 = \begin{bmatrix} 0 & 1 & 1 & 0 & 1 \\ 0 & 0 & 1 & 1 & 0 \\ 0 & 0 & 0 & 1 & 1 \\ 1 & 0 & 0 & 0 & 1 \\ 0 & 1 & 0 & 0 & 0 \end{bmatrix}$$

Now A_3 and A_4 differ non-trivially, for in A_3 there is an arc from the node of lowest out-degree to the node of highest out-degree, and this is not the case in A_4. (Davis (1954) discusses how to enumerate the distinct dominance patterns which have identical score sequence.)

The second step is to construct the **hierarchy index** h from the components of V. This index is defined as the variance of the V_i, normalized so that $h = 1$ in the fully hierarchical case. Clearly h is zero in the fully democratic case when that can be attained. The expression used is

$$h = \frac{12}{(N-1)N(N+1)} \sum_i [V_i - \tfrac{1}{2}(N-1)]^2$$

For even N, the lowest attainable h is $3/(N^2 - 1)$. The ambiguous score sequence vector discussed above has $h = 0.2$, and the example cited from Schjelderup-Ebbe has $h = 0.975$.

The subset of seven species already mentioned, in the marine community studied by Buss and Jackson (1979), display an adjacency matrix

$$A_5 = \begin{bmatrix} 0 & 1 & 1 & 0 & 1 & 1 & 1 \\ 0 & 0 & 1 & 1 & 0 & 1 & 0 \\ 0 & 0 & 0 & 0 & 1 & 1 & 1 \\ 1 & 0 & 1 & 0 & 1 & 1 & 1 \\ 0 & 1 & 0 & 0 & 0 & 1 & 0 \\ 0 & 0 & 0 & 0 & 0 & 0 & 1 \\ 0 & 0 & 0 & 1 & 1 & 0 & 0 \end{bmatrix}$$

with score sequence vector $V_5 = [5, 5, 3, 3, 2, 2, 1]$ and hierarchy index $h = 0.5$.

It is obvious that the choice of hierarchy index is bound to be to some extent arbitrary. This one is identical with an index adopted by statisticians in the context of preferences between pairs, such as consumer preferences of brand X over brand Y. The quantity $1 - h$ is proportional to the number of 3-cycles [i dominates j, j dominates k, k dominates i]. This helps us to see how information is lost going from the score sequence vector to the hierarchy index. Here are two adjacency matrices which both have two 3-cycles, while the second also has a 4-cycle, $3 \to 4 \to 5 \to 6 \to 3$:

$$A_6 = \begin{bmatrix} 0 & 1 & 0 & 1 & 1 & 1 \\ 0 & 0 & 1 & 1 & 1 & 1 \\ 1 & 0 & 0 & 1 & 1 & 1 \\ 0 & 0 & 0 & 0 & 1 & 0 \\ 0 & 0 & 0 & 0 & 0 & 1 \\ 0 & 0 & 0 & 1 & 0 & 0 \end{bmatrix}, \quad A_7 = \begin{bmatrix} 0 & 1 & 1 & 1 & 1 & 1 \\ 0 & 0 & 1 & 1 & 1 & 1 \\ 0 & 0 & 0 & 1 & 0 & 0 \\ 0 & 0 & 0 & 0 & 1 & 0 \\ 0 & 0 & 1 & 0 & 0 & 1 \\ 0 & 0 & 1 & 1 & 0 & 0 \end{bmatrix}$$

They have distinct score sequence vectors

$$V_6 = [4, 4, 4, 1, 1, 1], \quad V_7 = [5, 4, 2, 2, 1, 1]$$

but in each case the hierarchy index h is 0.77. (Petraitis (1979) discusses additional indices which take account of the presence of higher cycles.)

2.2 A MODEL OF INTRINSIC TRAITS

One can imagine various mechanisms for establishing the relative status of members of a flock. On the one hand intrinsic traits such as weight or (in hens) comb size may be significant, dominance in an encounter going to the one scoring highest in some assessment of these. On the other hand daughters may adopt the status of their mother, as noted by Schulman and Chapais (1980) in the macaque monkey. A newcomer to a flock may adopt a status determined by her first few encounters chancing to be with individuals of high or low status (McBride 1958).

In his first paper Landau assumes that each member i of the flock is characterized once and for all by an ability vector

$$\boldsymbol{X}_i = [x_i^1, x_i^2, \ldots, x_i^r]$$

where the x_i^p measure individual characteristics which can contribute to dominance. These are assigned values according to some probability distribution. Ranks are assumed to be established by a round robin of encounters between all possible pairs, which have not previously been in contact. In any encounter of i and j the probability P_{ij} that i should dominate j is assumed to be a function of $\boldsymbol{X}_i$ and $\boldsymbol{X}_j$. The significance of h in Landau's work is that he is able to set down values for the expected value and variance of this quantity, $E(h)$ and $\sigma'(h)$, directly in terms of these P_{ij}. So the hierarchical aspect of the pattern of dominance relations, for any particular model of this type, can be studied in terms of $E(h)$ and $\sigma'(h)$ without explicitly employing the adjacency matrices or the score sequence vectors. In an extreme case, in which $\boldsymbol{X}_i$ has only one component x_i and

$$P_{ij} = 1 \quad \text{if } x_i > x_j$$
$$= 0 \quad \text{if } x_i < x_j$$

the outcome must be strict hierarchy. Landau's results lead as they should to

$$E(h) = 1, \quad \sigma'(h) = 0$$

Adding independent x_i^p, and blurring the sharp transition in the P_{ij}, leads to

$$E(h) \ll 1$$

This result confirms the intuition that democratic rather than hierarchic patterns are likely when numerous intrinsic traits, distributed at random among the flock, determine the outcomes of the pairwise encounters.

Landau (1968) gives some information conflicting with the model of intrinsic traits. In one experiment (unfortunately he gives no reference) male sex hormone was injected into some of the lowest ranking hens. Their status rose, suggesting that hormone balance is a relevant trait. However, their status stayed at its new higher level after the physiological effects of the hormone wore off. This suggests that these hens, and the others, learned the new status and behaved so as to maintain it. A somewhat similar conclusion is suggested by Landau's other example, which relates to work done by King (1965). He established three successive pecking orders in two groups of ten hens. Order 1 was established by staged encounters between hens kept in isolation between encounters. Order 2 was established by keeping the hens isolated for a few weeks and then having them spend some time as a flock with ten members. After another interval, order 3 was established in the same manner as order 1. There was essentially no correlation between the 'naive' order 1 and the other two orders.

2.3 THE DYNAMIC INTERACTION DIGRAPH

How then should one address the problem of the observed preponderance of hierarchy? The history of the flock is probably relevant, and members may well have some awareness of their rank relative to others, in the form of memories of the outcome of recent encounters. Landau's second paper examines hypotheses of this kind, applied to a picture of the dynamics of the flock as a Markov process. In this picture one must define the current state of the flock, in some manner incorporating the current pattern of dominance relations, and also define how this state can change as a result of local piecemeal changes which come about by the reversal of individual dominance relations in a sequence of pairwise encounters. For each two individuals *i* and *j* with current scores V_i and V_j one may take the probability of a reversal ['*i* dominates *j*' switching to '*j* dominates *i*'] as a function of V_i and V_j. Landau emphasizes that not even the score sequence vector, let alone the hierarchy index, suffices to describe the state of the flock. The full adjacency matrix must be used.

Consider, for example, the ambiguous score sequence vector already examined

$$V = [3, 2, 2, 2, 1]$$

which can correspond to either A_3 or A_4. An encounter between hen 1 and hen 5, with dominance reversal, can lead from a state described by

$$A_3 = \begin{bmatrix} 0 & 1 & 1 & 1 & 0 \\ 0 & 0 & 0 & 1 & 1 \\ 0 & 1 & 0 & 0 & 1 \\ 0 & 0 & 1 & 0 & 1 \\ 1 & 0 & 0 & 0 & 0 \end{bmatrix}$$

to one described by

$$A_7 = \begin{bmatrix} 0 & 1 & 1 & 1 & 1 \\ 0 & 0 & 0 & 1 & 1 \\ 0 & 1 & 0 & 0 & 1 \\ 0 & 0 & 1 & 0 & 1 \\ 0 & 0 & 0 & 0 & 0 \end{bmatrix}$$

with score sequence vector [4, 2, 2, 2, 0]. On the other hand, no single reversal can lead from

$$A_4 = \begin{bmatrix} 0 & 1 & 1 & 0 & 1 \\ 0 & 0 & 1 & 1 & 0 \\ 0 & 0 & 0 & 1 & 1 \\ 1 & 0 & 0 & 0 & 1 \\ 0 & 1 & 0 & 0 & 0 \end{bmatrix}$$

to a state having this score sequence vector.

The Markov process can be thought of in terms of a **dynamic interaction digraph**, which can be set up in the following way. Label all members of the flock 1 to N in an arbitrary order. Label all interactions in lexicographic order, (1, 2), (1, 3), . . . , (1, N), (2, 3), (2, 4), . . . , ($N-1$, N). There are $M = \frac{1}{2}N(N-1)$ of these interactions. Assign to each of these interactions (i, j) the state 1 if i dominates j, 0 if j dominates i. The state of the system is now a vector of dimension M with components 1 or 0. This can be identified with a vertex of a hypercube in M dimensions. This is an appropriate representation not just because there are the correct number 2^M of vertices, but because each vertex has as its neighbours those corresponding to the states which can be reached by a single reversal. So the state transitions are along the edges of the hypercube, and the model assigns the relative probability of moving one way or the other along one of these edges. The Markov process consists of starting at an arbitrary vertex, choosing which edge is available (which two members meet next), deciding whether to move along this edge (whether a reversal occurs), and repeating this procedure at the new vertex.

Let us take as an example the rather trivial one of a flock of three members for which there are only three interactions, (1, 2), (1, 3), and (2, 3). There are eight states

000 in which 2 beats 1, 3 beats 1 and 3 beats 2

100 in which 1 beats 2, 3 beats 1 and 3 beats 2

. . . .

111 in which 1 beats 2, 1 beats 3 and 2 beats 3

Six states are hierarchic and only two, 010 and 101, are democratic. These eight states can be arranged as shown in Fig. 2.2 as the vertices of a cube. A reversal is a change, from 0 to 1 or from 1 to 0, of one of the three digits which define the state. This corresponds to moving from a vertex to one of its three neighbours. The Markov process is started by picking an initial state, say 000, and one of the digits, say the second. So the first encounter is between the 'weakest' member and the 'strongest', 1 and 3 respectively. The particular model to be tested will specify

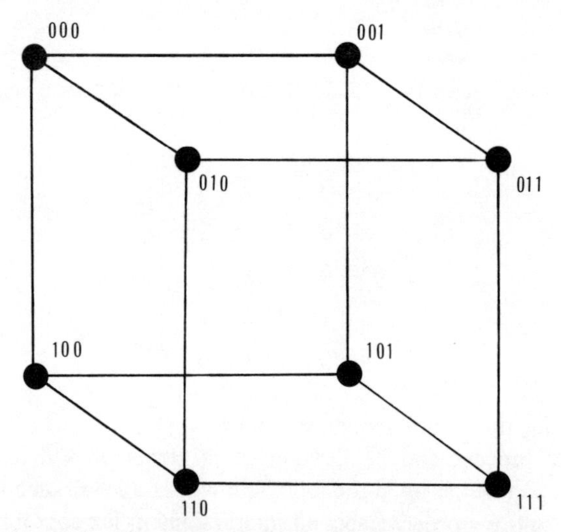

Fig. 2.2 The states of a flock of three members (and consequently with three pairwise interactions) shown as the vertices of a cube. The vertices 010 and 101 are democratic states, the others are hierarchic. This predominance of hierarchic states is a result of the small number of members

the relative probabilities that this encounter will result in a reversal, so that the new vertex is 010, or that it will not, leaving the vertex as 000. Continuing in this way there evolves a certain probability distribution for occupancy of the eight vertices. If this favours hierarchic vertices the model is consistent with the observed preponderance of hierarchy.

An example of a model which reinforces existing status differences, and which Landau found favoured hierarchic vertices, takes the following set of probabilities: 'i beats j' changes to 'j beats i' with probability $p(V_i - V_j)$, where $p = 1$ for $V_i - V_j \leqslant -2$, $p = 0$ for $V_i - V_j \geqslant 2$, and $0 < p < 1$ otherwise.

References

Buss, L. W., and Jackson, J. B. C. (1979). Competitive networks: intransitive competitive relationships in cryptic coral reef environments, *Amer. Nat.* **113**, 223–234.

Cole, B. J. (1981). Dominance hierarchies in *leptothorax* ants, *Science* **212**, 83–84.

Davis, R. L. (1954). Structures of dominance relations, *Bull. Math. Biophys.* **16**, 131–140.

Gurney, W. S. C. and Nisbet, R. M. (1979). Ecological stability and social hierarchy, *Theor. Popn. Biol.* **16**, 48–80.

King, M. G. (1965). The effect of social context on dominance capacity of domestic hens, *Animal Behaviour* **13**, 132–133.

Landau, H. G. (1951a). On dominance relations and the structure of animal societies—I Effect of inherent characteristics, *Bull. Math. Biophys.* **13**, 1–19.

Landau, H. G. (1951b). On dominance relations and the structure of animal societies—II Some effects of possible social factors, *Bull. Math. Biophys.* **13**, 245–262.

Landau, H. G. (1968). Models of social structure, *Bull. Math. Biophys.* **30**, 215–224.

Lomnicki, A. (1978). Individual differences between animals and the natural regulation of their numbers, *J. Anim. Ecol.* **47**, 461–476.

McBride, G. (1958). Measurement of aggressiveness in the domestic hen, *Animal Behaviour* **6**, 87–91.

Petraitis, P. S. (1979). Competitive networks and measures of intransitivity, *Amer. Nat.* **114**, 921–925.

Schjelderup-Ebbe, T. (1922). Beitrage zum Sozial-psychologie des Haushuhns, *Zeit fur Psychol.* **88**, 225–252.

Schulman, S. R., and Chapais, B. (1980). Reproductive value and rank relations among macaque sisters, *Amer. Nat.* **115**, 580–595.

Chapter 3

The trophic structure of foodwebs

The main subject of the next four chapters is the structure of communities of organisms. The components studied are populations of animals and plants which share a common habitat. The interactions of such populations can be diverse, including predation, parasitism, competition, and mutualism, but the emphasis here is mainly on predation. In this chapter only predation is considered, while in the following chapter a particular form of competition is looked at, which is the result of two species having a common prey. The sets of data mainly studied in Chapters 3 and 4 are of the type usually termed a foodweb. This consists of the enumeration of certain components and of the relationships between them of the form 'i eats j'. Neither the relative abundances of these components, nor the contribution which j makes to the calorific intake of i, is specified.

Even within this very bare description of the community, webs are unlikely to be complete, and there can be ambiguities in assigning the components and identifying the relationships. In principle species can be distinguished, and one might prefer to carry out the analysis at species level. However, in many of the available foodwebs species have been lumped together so that a component may be 'grass', 'spiders', and so on. Worse still, in some of the webs some components are a single species but others are lumped. In order to avoid the appearance of loops, 'i eats i', and cycles, 'i eats j, j eats i', species may on the other hand be split up between different components, such as adults and young. In general I shall assume that this can be done and that no foodwebs contain loops or cycles. It is clear that foodwebs are studied in default of more detailed data on the quantitative aspects of predation. Later in this chapter I shall discuss what can be done in the relatively rare cases where such data are available for moderately large sets of components. I shall call these cases weighted foodwebs.

In the introduction to his compilation and analysis of foodwebs Cohen (1978) draws some distinctions between different types of foodweb. Similar foodwebs are likely to be found in distinct but similar habitats, and sometimes data are presented which cover more than one habitat. Cohen therefore distinguishes between single habitat webs and multi-habitat webs. In some cases the components present have been identified and their interactions traced out. Cohen calls this a community web. In others the investigation has started from a single plant species or some other source of food, and then components eating this are identified, and then those eating this set of components, ... Cohen calls this a source web. Finally the starting point may be a single carnivore species or a wider category of carnivores, and their prey and the prey of these, . . . identified. Cohen calls this a sink web. The implication of these distinctions is that one would prefer to use only single-habitat community webs, and the others are used because of the lack of data.

3.1 TROPHIC IMPURITY

The most striking structural feature of foodwebs is their division into trophic levels. Components can usually be identified as plants, herbivores, carnivores, carnivores on carnivores, etc., with sometimes up to five or six distinct levels. Closer examination reveals quite a few cases in which a species obtains its food from more than one lower level, such as a bird which eats seeds and insects, or a large fish which eats both carnivorous and non-carnivorous small fish. But it is usually reasonable to treat the web as basically a set of trophic levels with occasional variations. Most of the available sets of foodweb data do not contain information about reducers such as bacteria which close the nutrient loop by returning biomass to a form available to plants.

So in setting down digraphs for foodwebs, with nodes for components and arcs from prey to predator, I shall start from the idealized case of a strictly trophic web. An example is given in Fig. 1.2. The nodes are divided into P sets I_p such that

if i, j are in the set I_p there is no arc (i, j) or (j, i)
if i and j are respectively in sets I_p and I_q with $|p-q| \geqslant 2$, there is no arc (i, j) or (j, i)
if i, j are respectively in sets I_p and I_{p+1}, there can be an arc (i, j) but not an arc (j, i)

A concrete example is necessary to show some of the consequences of interactions which skip a trophic level. I present in Fig. 3.1 a foodweb from the compilation of Cohen (1978). The original source of the data is Minshall (1967). This is a freshwater community web with thirteen components, all of which are lumped, not individual, species. There are five top predators:

1. Phagacota
2. Decapoda
3. Plecoptera
4. Megaloptera
5. Fish

There are two source components:

12. Detritus
13. Diatoms

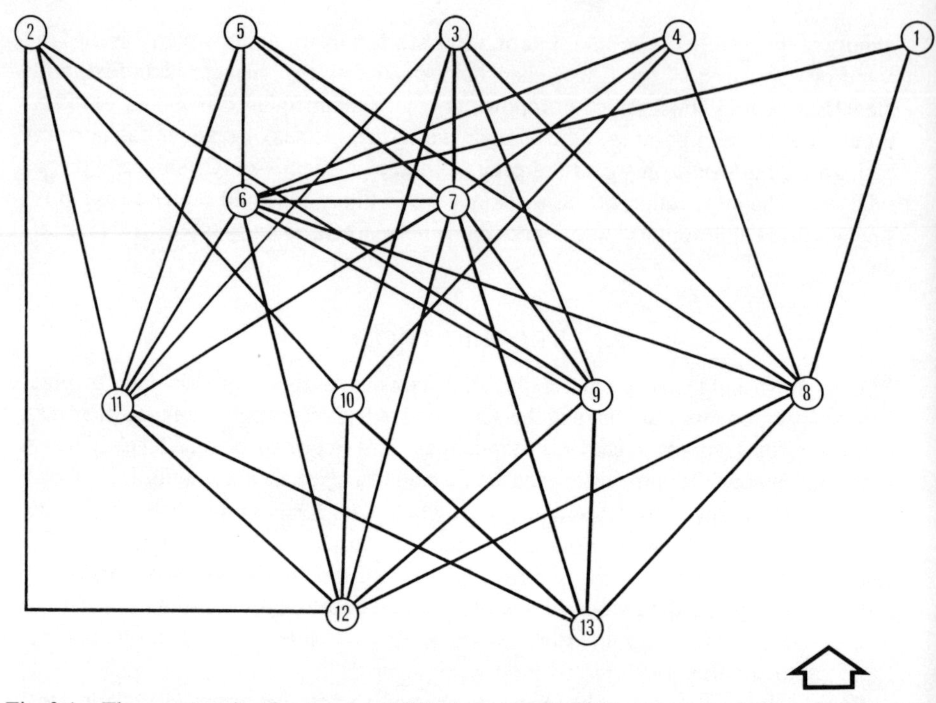

Fig. 3.1 The community food web of Minshall (1967) displayed as a digraph. Arrows are omitted; they point in the direction of the large arrow, and also from 7 to 6. The components of the web are numbered as in the text

There are four components that feed only on components 12 and 13, making a clearly defined second trophic level:

8. *Asellus*
9. Ephemeroptera
10. Trichoptera
11. Diptera

The trophic status of two components is unclear:

6. *Gammerus*
7. other Trichoptera

Even for this fairly small web the pictorial representation of the digraph is rather confusing, although some of the features just mentioned can be brought out by plotting 1 to 5, 8 to 11, and 12 and 13 in parallel rows. It is best to make use of the adjacency matrix, presented in Table 3.1. It is convenient to suppress two blocks of zeros which are always present in the full adjacency matrix, because by definition nothing eats a top predator and source components eat nothing. The rows are labelled by the numbers identifying components that are eaten and the columns by the numbers identifying components that eat. The numbers 6 to 11 show up in both rows and columns.

Table 3.1 Adjacency matrix for the food web shown in Fig. 3.1. A 1 in the column labelled i and the row labelled j means that the component i of the web eats the component j

	1	2	3	4	5	6	7	8	9	10	11
6	1	0	0	1	1	0	0	0	0	0	0
7	0	0	1	1	1	1	0	0	0	0	0
8	1	0	1	1	1	1	0	0	0	0	0
9	0	1	1	0	0	1	1	0	0	0	0
10	0	1	1	1	0	0	0	0	0	0	0
11	0	1	1	0	1	1	1	0	0	0	0
12	0	1	0	0	0	1	1	1	1	1	1
13	0	0	0	0	0	0	1	1	1	1	1

Reproduced from *Food Webs and Niche Space*, by J. E. Cohen, by permission of the Princeton University Press. © 1978 by the Princeton University Press

It would be convenient to describe this foodweb as a web of P levels to an accuracy α, knowing that the value chosen for P minimizes α. If P is taken as 5, the obvious allocation of the components is

I_1 contains 12, 13; I_2 contains 8 to 11; I_3 contains 7;
I_4 contains 6; I_5 contains 1 to 5

Then out of the total number of arcs $A = 36$ in the digraph, 23 skip a level, and it is natural to take $\alpha = 23/36$. If P is taken as 4, by putting 6 as well as 7 in I_3, and 1 to 5 in I_4, then the number of arcs that skip a level falls to sixteen, but there is also an arc, from 7 to 6, within level I_3. So $\alpha = 17/36$. Finally the assignment of $P = 3$, with components 6 to 11 in I_2 and 1 to 5 in I_3, gives only one arc skipping level I_2, but seven within I_2. So this yields the smallest α, equal to 2/9. A suitable name for this index α is trophic impurity.

As well as the 'horizontal' trophic levels, one can count 'vertical' food chains, each with one component in I_1 and one in I_p, with up to $P-2$ intermediate components. An example here is the chain 12, 9, 7, 4. There are 64 of these chains in this web. It is normally closer to the truth to describe a foodweb as having several trophic levels, with occasional arcs missing out a level, than as several parallel chains with occasional cross-links. I shall discuss later the reasons for adopting the ratio of the number of arcs A to the number of nodes N as an index of the overall complexity of a foodweb. In the webs included in Cohen's compilation this quantity has an average value of about 1.9, with extreme values 1.2 and 3.0. In this particular web the ratio is about 2.8, so it is a relatively complex web (MacDonald 1979).

3.2 REDUCTION OF A WEIGHTED FOODWEB TO A FOOD CHAIN

Given more complete data, including estimates of the energy transfers between components, one may still wish to reduce the weighted foodweb to a simple sequence of trophic levels. Rather than assigning each component to a single trophic level, and trying various assignments to see which works best, it is now possible either to partition the component between two or more levels, or to assign it a fractional trophic position. Consider a case in which the lowest trophic level and the next one up are unambiguously identified, and in which there is a component that obtains half its energy intake from each of these two levels. Then one can either say that levels 2 and 3 each contain a half share of this component, or one can say that this component has trophic position 2.5. I shall examine in more detail two particular methods that have been developed, each from one of these viewpoints.

It is necessary to count, and to weight appropriately, the alternative paths which energy follows from the primary producers to the higher components. Path counting is conveniently done using the adjacency matrix of the digraph; indeed the primary use of this matrix in graph theory is for counting paths. Recall that the adjacency matrix A is defined by

$$A_{ij} = 1, \text{ if there is an arc } (i,j)$$
$$A_{ij} = 0, \text{ otherwise}$$

Recall also that I have excluded loops (i, i) from foodweb digraphs, and that matrix multiplication is defined in such a way that the elements of the matrix A^2 are given by

$$A^2_{ik} = \sum_j A_{ij} A_{jk}$$

Then it is clear that

$$\begin{aligned}[A^2]_{ik} &= 1, \text{ if there is an arc } (i,j) \text{ and an arc } (j,k) \text{ for only one node } j \\ &= 2, \text{ if there are arcs } (i,j), (i,j'), (j,k) \text{ and } (j',k) \text{ for two distinct nodes } j, j', \\ &\quad \text{and so on}\end{aligned}$$

So elements of A^2 count the number of paths of length 2 from one node to another. In the same way, elements of A^r count the number of paths of length r from one node to another.

Naturally if paths have to be weighted things become more complicated. Also it is necessary to take account of inputs; usually only of solar energy at the lowest level but sometimes also at higher levels, for example, if the habitat is a river pool and some organisms flow in with the current. It is necessary to take account of outputs, for example downstream in the case of the river pool, or to man in a harvested foodweb. Most important of all, it is necessary to take account of respiration, which is unusable output. For any component the first law of thermodynamics tells us that input plus transfers from other components must equal output plus respiration plus transfers to other components. The second law tells that respiration must be present. So for each component there is an energy balance equation

$$\sum_{j \neq i} P_{ji} + E_i = \sum_{k \neq i} P_{ik} + R_i + X_i \tag{3.1}$$

with $R_i > 0$. Here E_i, R_i, and X_i are input, respiration, and output respectively, and P_{ij} is energy transfer from i to j.

Ulanowicz and Kemp (1979) address themselves to the question how best to reduce a foodweb from its original N components to a new set of M 'blended' components displaying a pure trophic structure. They emphasize that the reduction should conserve the total input, respiration, and output of the web, and should yield positive values of these quantities for each of the new components. In order to weight, rather than simply to count, the paths, they replace the adjacency matrix A by a matrix of feeding coefficients G

$$G_{ij} = \frac{P_{ij}}{\sum\limits_{k \neq j} P_{kj} + E_j} \tag{3.2}$$

This matrix element is the fraction of the energy intake of component j that comes directly from component i. Since G has a zero element wherever A has a zero element, it follows that $[G^2]_{ij}$ gives the fraction of the energy intake of j which flows from i by paths of length 2, and so on. One has also to define the fraction of the energy intake of j that comes from outside the web, as

$$F_j = \frac{E_j}{\sum\limits_{i \neq j} P_{ij} + E_j} \tag{3.3}$$

Let F be the column vector with elements F_j and F' the corresponding row vector. Then the elements of the row vector $t_2 = F'G$ are the fractions of the energy intakes of the various components which come from outside the web by way of one intermediate component. In the same way define $t_3 = F'G^2$, and so on, and let $t_1 = F'$. Ulanowicz and Kemp transform the original matrix P with elements P_{ij}, which is an $N \times N$ matrix, into a new matrix Q which is $M \times M$, where M is the longest path length from any component of the web to an outside energy source. The transformation is

$$Q = T'PT \tag{3.4}$$

where the rows of the $M \times N$ matrix T' are the row vectors t_1 to t_M. They prove that the column sums of T' are all equal to 1, and that this guarantees the conservation of the total energy input, total respiration, and total usable output.

At this stage one has obtained an acceptable compression of the original set of N components into a new set of M components, which has in it paths as long as any in the original web, and a correct balance of paths of different lengths. It is by no means a food chain. Tables 3.2 and 3.3 show the data for a North Sea foodweb, adapted by Ulanowicz and Kemp from work of Steele (1974), before and after this transformation from $N = 11$ to $M = 8$. The transformed matrix resembles a chain in that elements Q_{ii+s} fall off strongly with s. But the components at higher trophic levels receive a large proportion of their energy intake from components other than their immediate predecessor. Ulanowicz and Kemp describe a method for obtaining an equivalent 'canonical' food chain by

Table 3.2 Energy transfers, in kcal/m^2 per year, in a North Sea foodweb. *I* is input (solar energy), *R* is respiration, and *O* is output (for example, to man). A pure food chain would have entries only in these columns and in positions such as row 1, column 2. The components of web are: (1) primary producers, (2) pelagic herbivores, (3) bacteria, (4) meiobenthos, (5) macrobenthos, (6) invertebrate carnivores, (7) pelagic fish, (8) other carnivores, (9) demersal fish, (10) large fish

	I	*R*	1	2	3	4	5	6	7	8	9	10	*O*
1	9000	8100		900									
2		430			300			85	85				
3		179				21	100						
4		1					20						
5		70								20	30		
6		74							11				
7		88									4		4
8		18									2		
9		33										0.6	2
10		0.54											0.06

Reproduced from 'Towards canonical trophic aggregations', *Amer. Nat.* **114**, 871–883, by R. E. Ulanowicz and W. M. Kemp, by permission of the University of Chicago Press. © 1979 by the University of Chicago.

Table 3.3 Energy transfers, in kcal/m^2 per year, in the web of Table 3.2, when transformed in the first stage of the method of Ulanowicz and Kemp. The rows and columns are labelled to indicate successively higher levels, but some of the higher levels do not receive most of their input from the level immediately below. *I*, *R*, and *O* have the same meanings as in Table 3.2

	I	*R*	1	2	3	4	5	6	7	8	*O*
1	9100	8000		900							
2		430			460	9.74					
3		331				106	19.17	0.66	0.33		3.54
4		72.7					34.6	7.50	0.25	0.00055	0.655
5		50.3						1.84	0.14	0.0039	1.42
6		9.57							0.02	0.001	0.03
7		0.41								5×10^{-4}	0.03
8		0.005									5.6×10^{-4}

Reproduced from ‘Towards canonical trophic aggregations’, *Amer. Nat.* **114**, 871–883, by R. E. Ulanowicz and W. R. Kemp, by permission of the University of Chicago Press.

successively eliminating all Q_{ij} with $j > i + 1$. They emphasize that this second state does not guarantee the preservation of the first and second laws. The results of this second transformation are shown in Table 3.4.

Merely to apply these methods to a single example tells one little; they are of interest only if they can be used to compare the structure of different foodwebs. Ulanowicz (1982) has applied the canonical chain method as part of a study of two marsh creek communities in Florida. (This study uses unpublished data of M. Homer and W. Kemp, presented in part by Ulanowicz (1980).) The two communities have about twenty components, with some lumping of species at low trophic levels. The flows include cycling by way of detritus, detrivores, and bacteria. Ulanowicz emphasizes that separation of cyclic and non-cyclic flows, a necessary preliminary to using the canonical chain method, is an ambiguous process. After removal of a suitable set of cyclic flows there remains a foodweb with the residual flows between components, respiration, and outputs. This can now be reduced to a canonical food chain. Two creeks are studied, one as a control regarded as typical of such creeks in the region, the other subject to 6°C average temperature elevation caused by effluent from the Crystal River nuclear generating station. Table 3.5 shows the two canonical chains, each of which has five trophic levels. Flow to level 5 is reduced by a factor of ten in the perturbed community relative to its value in the control community. This corresponds, in the original web, to a reduction in the variety of prey consumed by some of the fish species.

When quantitative data on cycles are available it seems that something more constructive should be done with them than in this procedure. In fact much of Ulanowicz' recent work centres on the use of descriptive indices appropriate to webs which include cycles. The analysis of webs with cycles will be the subject of a forthcoming monograph by Ulanowicz.

The complementary viewpoint to the canonical chain method, in which a fractional trophic position is assigned to each component, is the subject of a paper by Levine (1980). Let there be an external input feeding only the lowest trophic level, so that there are components with trophic position 1. Let component i be connected to this source by paths of lengths k running from a minimum value $k(i)$ to a maximum value $K(i)$, and weighted, using a G matrix as in the work of

Table 3.4 Energy transfers for the canonical food chain of the North Sea foodweb of Table 3.2. The units are kcal/m^2 per year. For each level the entries in the rows O, T, and R stand for output (for example, to man), transfer to the next level up, and respiration respectively

	1	2	3	4	5	6	7	8
O			3.63	0.98	1.36	0.095	0.00124	-2.3×10^{-4}
T	900	470	127	40.4	0.574	0.00158	10^{-5}	
R	8100	430	339	85.6	38.5	0.478	3.3×10^{-4}	2.4×10^{-4}

Reproduced from 'Towards canonical trophic aggregations', *Amer. Nat.* **114**, 871–883, by R. E. Ulanowicz and W. M. Kemp, by permission of the University of Chicago Press.

Table 3.5 Energy transfers, in kcal/m^2 per year, for the canonical food chains, constructed by the method of Ulanowicz and Kemp, for unstressed (**A**) and stressed (**B**) communities in Florida creeks. The columns and rows are identified as in Table 3.4. The inputs to level 1 are 7358 in **A** and 6018 in **B**

A	1	2	3	4	5
O	219	592	448	4.68	0.88
T	4622	1475	22.4	3.15	
R	2517	2555	1004	14.6	2.27
B					
O	167	634	50.2	2.45	0.072
T	3520	794	13.8	0.29	
R	2331	2092	730	11.1	0.213

Reproduced from 'Identifying the structure of cycling in eco-systems', Ecology (forthcoming) by R. E. Ulanowicz, by permission of the Duke University Press. © 1982 by the Ecological Society of America.

Ulanowicz and Kemp, by factors $W(i, k)$. Then the trophic position of i is defined to be

$$x_i = \sum_{k=k(i)}^{K(i)} k\, W(i, k) \tag{3.5}$$

A measure of the trophic spread of the component i is the variance

$$\sigma_i = \sum_{k=k(i)}^{K(i)} (k - x_i)^2\, W(i, k) \tag{3.6}$$

Table 3.6 shows these quantities for the 10 components of the same North Sea web as before, as well as an additional component, man. A measure of the trophic impurity of a web could be the average value of σ_i over all i. For this North Sea web, excluding man, this comes to 0.22.

Table 3.6 Trophic positions, obtained by the method of Levine, for the components of the North Sea foodweb of Table 3.2. The components are labelled as in Table 3.2, with the addition of man, component 11. The mean trophic position of a component is x, and its variance is σ

	1	2	3	4	5	6	7	8	9	10	11
x	1	2	3	4	4.17	3	3.12	5.17	5.11	6.11	4.8
σ	0	0	0	0	0.37	0	0.32	0.37	0.55	0.55	1.04

Reproduced from 'Several measures of trophic structure applicable to complex food webs', *J. theor. Biol.* **83**, 195–207, by S. H. Levine, by permission of Academic Press Inc. (London) Ltd. and the author. © 1980 by Academic Press.

Table 3.7 Trophic positions x, and their variances σ, for the foodweb of Fig. 3.1 and Table 3.1. The components are labelled as in the text. It is assumed that a component with n prey obtains its energy input equally from each of them

	1	2	3	4	5	6	7	8	9	10	11	12	13
x	4.06	2.86	3.22	3.81	3.81	3.13	2.67	2	2	2	2	1	1
σ	0.41	0.12	0.17	0.45	0.45	0.29	0.22	0	0	0	0	0	0

It is interesting to apply this approach to the web studied earlier in this chapter, with adjacency matrix set out in Table 3.1. For this we had an estimate 2/9 for the trophic impurity, obtained by the assumption of a three-level structure. By direct counting or by using powers of the adjacency matrix, we arrive at the results set out in Table 3.7, for trophic positions and variances of the thirteen components. These in turn yield an estimate of 0.16 for the trophic impurity of the whole web.

References

Cohen, J. E. (1978). *Food Webs and Niche Space*, Princeton University Press, Princeton.

Levine, S. H. (1980) Several measures of trophic structure applicable to complex foodwebs, *J. theor. Biol.* **83**, 195–208.

MacDonald, N. (1979). Simple aspects of foodweb complexity, *J. theor. Biol.* **80**, 577–588.

Minshall, G. W. (1967). Role of an allochthonous detritus in the trophic structure of a woodland springbrook community, *Ecology* **48**, 139–149.

Steele, J. H. (1974). *The Structure of Marine Ecosystems* Harvard University Press, Cambridge.

Ulanowicz, R. E., and Kemp, W. M. (1979). Toward canonical trophic aggregations, *Amer. Nat.* **114**, 871–883.

Ulanowicz, R. E. (1980). An hypothesis on the development of natural communities, *J. theor. Biol.* **85**, 223–245.

Ulanowicz, R. E. (1982). Identifying the structure of cycling in ecosystems, submitted to *Ecology*.

Chapter 4

The interval structure of foodwebs

This chapter is concerned with the competition between components of a foodweb, which is present whenever two components share one or more prey. I shall first make some brief comments on relevant aspects of competition, and then introduce the consumer graph of a foodweb, which isolates the competitive aspects of the web, and a complementary resource graph. I shall discuss an abstract graph property, the interval property, which Cohen (1978) has found to be prevalent in the consumer graphs of empirical foodwebs. Finally I shall discuss the interpretations by Cohen (1978) and by Sugihara (1982) of this property in relation to the aspects of competition discussed at the beginning of the chapter.

4.1 CONTINUOUS AND DISCRETE RESOURCES

A basic difficulty in the formulation of a mathematical treatment of competition is that neither a continuous nor a discrete description of shared resources seems adequate. Let us consider two species which live in estuarine water, and each of which can only thrive within a certain range of salinity. If these ranges do not overlap the species are ecologically isolated and cannot be said to compete. If the ranges overlap and the species have other shared resources, the species compete. Now consider the food resources (prey species) available in the shared salinity range. Assume that there are five species present: *A*, *B*, *C*, *D*, and *E*. If predator species 1 eats *A* and *B*, and predator species 2 eats *C*, *D*, and *E*, they are ecologically isolated. If predator species 1 also eats *C* the species 1 and species 2 compete. Thus for competition we have a combination of the overlap of the suitable locations along a continuous salinity range and intersection of discrete sets of prey species. The usual discussion of niche overlap refers to a resource space defined by several continuous resource axes. It is important to note that if more than one type of resource is involved, *all* must overlap or intersect if the species are to be said to compete.

Let us now consider the addition of a third predator, differing in its range of acceptable salinity. It may be said to be added conservatively if the salinity ranges PQ and RS in Fig. 4.1 **A** are replaced by $P\dot{Q}$, $\dot{R}\dot{S}$ and TU in **B**. It may be said to

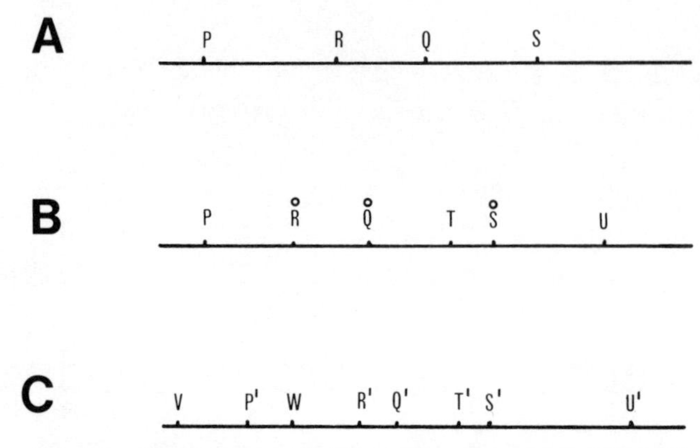

Fig. 4.1 The conservative (**B**) and non-conservative (**C**) addition of a new species to two which use overlapping intervals PQ and RS on a resource axis as shown in **A**. In **B** the range of the third species is TU, while in **C** it is the union of VW and T′U′

be added non-conservatively if the range of the new species is non-continuous, as indicated in C. Again consider the addition of a new species which finds its prey in the set A to F. Let us assume that these six prey species can be ordered in the sequence A to F, for example by size or by the trophic level to which they belong. Then species 3 may be said to be added conservatively if, for example, it eats E and F, or C to F, but non-conservatively if it eats A, E, and F. It may be noted that if competition simply means sharing one or more prey, then both the conservative case ABC, CDE, CDEF and the non-conservative case ABC, CDE, AEF involve competition of all three species.

Cohen's discussion of the interval property will be seen to rely on the continuous axis interpretation of a resource, and not to allow the conservative or non-conservative distinction. Sugihara, whose discussion is in terms of discrete resource sets, and who introduces the terms conservative and non-conservative (as well as consumer graph and resource graph) makes this distinction a crucial feature of his interpretation of foodweb structure.

4.2 THE CONSUMER GRAPH AND THE RESOURCE GRAPH

In this section I shall use the example of a foodweb that is illustrated in Fig. 3.1 and in Table 3.1. The **consumer graph** can be constructed from it as follows. First delete from the digraph any node with no arc leading to it (in this case nodes 12 and 13). Take the remaining nodes as the vertices of a graph. Put in edges ij whenever the digraph has at least one pair of arcs (k, i) and (k, j). This means that the i and j components of the foodweb share at least one prey. Finally (a step not needed in this particular case) any vertex not connected with any of the others should be discarded. The consumer graph is shown in Fig. 4.2 and its adjacency matrix in Table 4.1.

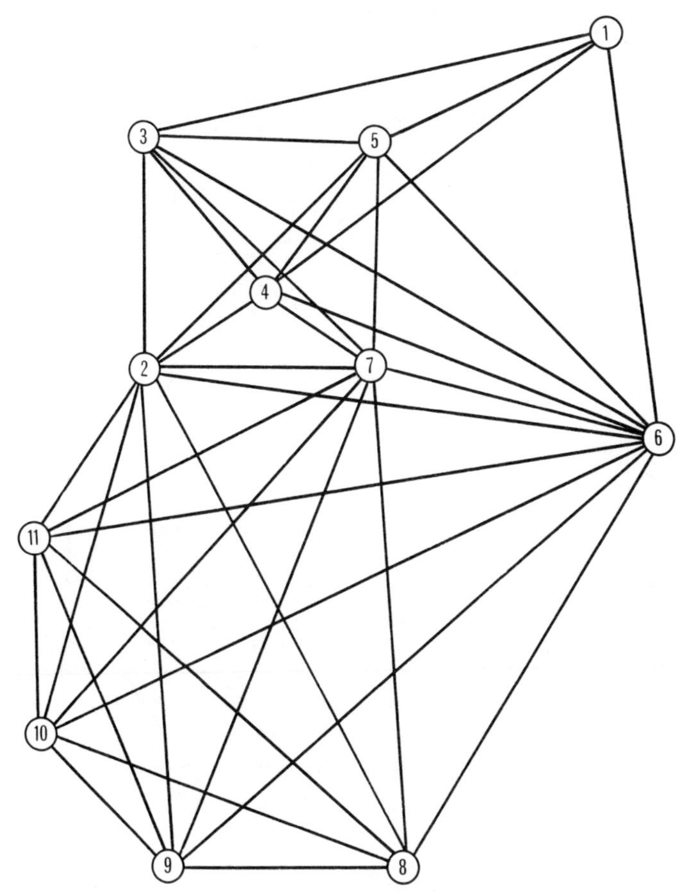

Fig. 4.2 The consumer graph constructed from the digraph of Fig. 3.1. Components 12 and 13 are absent, because they have no prey. The other components are numbered as before. The dominant cliques are (1,3,4,5,6), (2,3,4,5,6,7) and (2,6,7,8,9,10,11)

Table 4.1 The adjacency matrix for the consumer graph shown in Fig. 4.2. There is a high concentration of 1s, denoting competition, between members of the main trophic groupings of Fig. 3.1 and Table 3.1, namely components 1 to 5 and components 8 to 11

	1	2	3	4	5	6	7	8	9	10	11
1	0	0	1	1	1	1	0	0	0	0	0
2	0	0	1	1	1	1	0	1	1	1	1
3	1	1	0	1	1	1	0	0	0	0	0
4	1	1	1	0	1	1	0	0	0	0	0
5	1	1	1	1	0	1	0	0	0	0	0
6	1	1	1	1	1	0	1	1	1	1	1
7	0	0	0	0	0	1	0	1	1	1	1
8	0	1	0	0	0	1	1	0	1	1	1
9	0	1	0	0	0	1	1	1	0	1	1
10	0	1	0	0	0	1	1	1	1	0	1
11	0	1	0	0	0	1	1	1	1	1	0

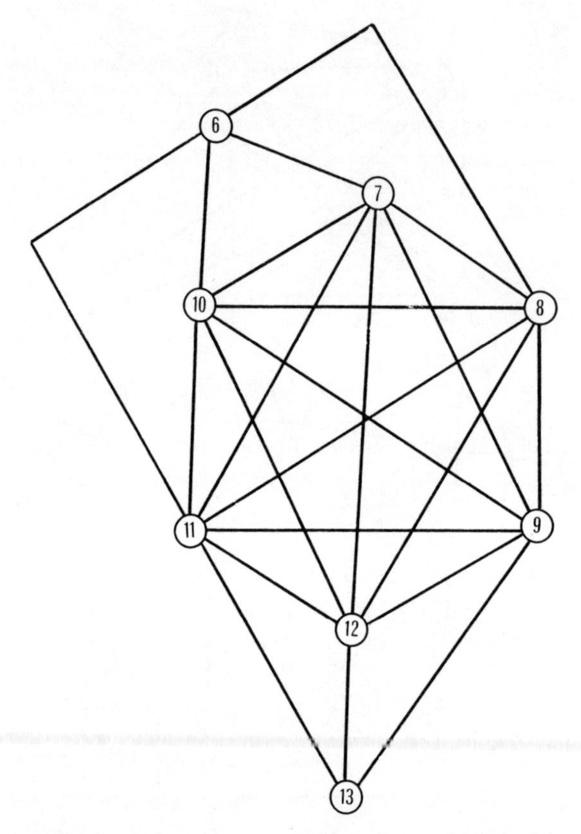

Fig. 4.3 The resource graph constructed from the digraph of Fig. 3.1. Components 1 to 5 are absent, because they are not the prey of any other components. The other components are numbered as before. The dominant cliques are (6,7,8,10), (9,11,12,13) and (7,8,9,10,11,12)

Table 4.2 The adjacency matrix for the resource graph shown in Fig. 4.3

	6	7	8	9	10	11	12	13
6	0	1	1	0	1	1	0	0
7	1	0	1	1	1	1	1	0
8	1	1	0	1	1	1	1	0
9	0	1	1	0	1	1	1	1
10	1	1	1	1	0	1	1	0
11	1	1	1	1	1	0	1	1
12	0	1	1	1	1	1	0	1
13	0	0	0	1	0	1	1	0

The **resource graph** is constructed in a complementary manner. First delete from the digraph any node with no arc leading from it (in this case nodes 1 to 5). Take the remaining nodes as the vertices of a graph. Put in an edge *ij* whenever the digraph has at least one pair of arcs (*i*, *k*) and (*j*, *k*). This means that in the foodweb the *i* and *j* components share at least one predator. Finally delete any vertex which is not connected to the remainder of the graph (again not necessary in this case). The resource graph is shown in Fig. 4.3 and its adjacency matrix in Table 4.2.

Cohen (1978) introduces the interval property only in relation to consumer graphs, but I have found that it is just as prevalent a property for resource graphs.

4.3 THE INTERVAL PROPERTY

There is an area of graph theory related to the interpretation of a vertex as a set or a hypervolume and an edge as the (non-empty) intersection of two sets or the overlap of two hypervolumes. Particular types of graph can be defined according to the nature of the interpretation adopted. In particular, an **interval graph** is one compatible with the interpretation of the vertices as intervals on the line and the edges as overlaps of such intervals. It is essential to note two points. First, given the original interpretation of a graph (such as the consumer graph), the fact that the graph is interval does not imply that any real meaning can be attached to positions along the line. Second, 'interval' does not imply that an interpretation in terms of a higher dimensional space is excluded.

Certain kinds of graph clearly cannot be interval. Two examples are shown in Fig. 4.4, along with illustrations of how they are interpreted by the overlap of regions in the plane. For a graph to be interval it must possess the **rigid circuit property**. This means that any polygon made up of four or more edges must possess a chord, that is an edge which shortens the path. The open square in Fig. 4.4**A** is the simplest example of a graph not having this property. In practice one can often readily identify a non-interval graph by locating in it a subgraph which is a non-rigid polygon, such as the square in Fig. 4.4**A**, or a **star** as shown in Fig. 4.4**B**. However (Gilmore and Hoffman 1964, Lekkerkerker and Boland 1962) there are some more complicated forbidden configurations, and in a very tangled graph the eye may miss even a square or a star, so a positive test for the interval property is needed (Fulkerson and Gross 1965).

A **clique** is a set of vertices with an edge linking every pair in the set. A **dominant clique** is a clique not contained in a larger clique. (In the special case of a foodweb with pure trophic structure, and with all pairs in every level except the lowest having shared prey, then in the consumer graph the dominant cliques are these trophic levels.) The dominant cliques can be set down in matrix form. This is done in Table 4.3 for the consumer graph and in Table 4.4 for the resource graph. In each of these examples we see the **consecutive ones property**, that in each column all ones appear in a set of consecutive rows. The result of Fulkerson and Gross (1965) is that a graph is interval if and only if it has this consecutive ones

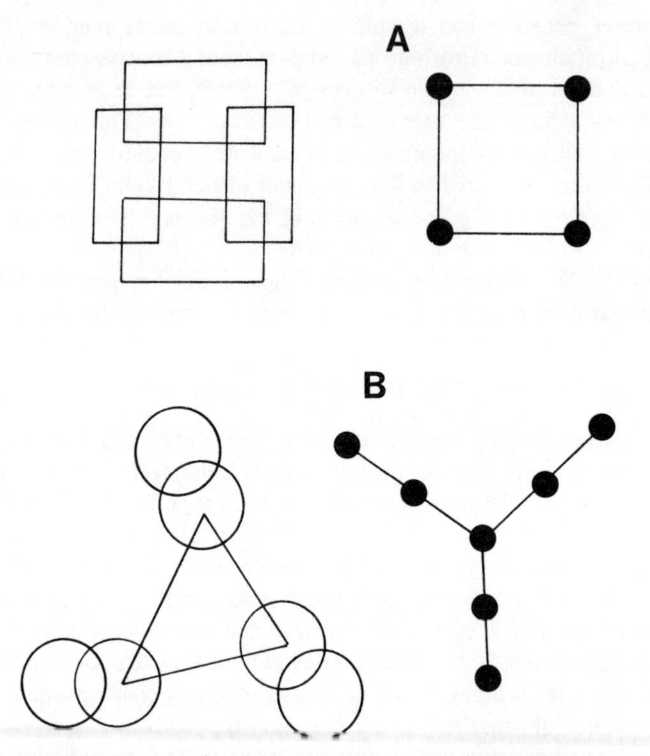

Fig. 4.4 The non-rigid square (**A**) and the star (**B**). These correspond to overlaps possible in two or more dimensions but not in one

Table 4.3 Dominant cliques for the consumer graph in Fig. 4.2 and Table 4.1, displaying the 'consecutive 1s' property. (Going down each column one encounters only consecutive 1s.) This property must hold when there are no more than three dominant cliques

1	2	3	4	5	6	7	8	9	10	11
0	1	0	0	0	1	1	1	1	1	1
0	1	1	1	1	1	0	0	0	0	0
1	0	1	1	1	1	0	0	0	0	0

Table 4.4 Dominant cliques of the resource graph of Fig. 4.3 and Table 4.2. Again there are only three, and the consecutive 1s property is trivial

6	7	8	9	10	11	12	13
1	1	1	0	1	0	0	0
0	1	1	1	1	1	1	0
0	0	0	1	0	1	1	1

Table 4.5 Dominant cliques of the open square (A) and of the star (B), which do not have the consecutive 1s property, and so are not interval

A	1	2	3	4
	1	1	0	0
	0	1	1	0
	0	0	1	1
	1	0	0	1

B	1	2	3	4	5	6	7
	1	1	0	0	0	0	0
	0	1	1	0	0	0	0
	0	0	1	1	0	0	0
	0	0	0	1	1	0	0
	0	0	1	0	0	1	0
	0	0	0	0	0	1	1

property. Table 4.5 shows the dominant cliques for the square and the star. No reordering of the rows will give this property to the columns.

An earlier application of interval analysis in biology was that by Benzer (1959) who investigated the fine structure of the gene in the virus phage T4. On the assumption that any mutation involves a localized portion of the gene, then by recombining two different mutant strains one can recover the original type, provided the two portions do not overlap. Benzer uses the illustrative analogy of two recorded tapes of a musical work, each with an error, such as a passage played incorrectly or obscured by noise. The correct version can be recovered provided the two erroneous passages do not overlap. Of course the tape is manifestly one-dimensional; the question that Benzer chose to investigate is whether this can be said of the gene. He was able to establish complete overlap data on one segment of gene giving rise to nineteen mutants, and incomplete data on a total of 145 mutants. He applied a procedure which amounted to the consecutive ones test of Fulkerson and Gross, to show that all these data are consistent with a one-dimensional fine structure of this gene. He neither used the term graph nor cited the mathematical literature. Indeed his work preceded the three primary mathematical publications in this area.

Cohen (1978) calls a foodweb 'interval' if its consumer graph possesses the interval property. I shall call a foodweb 'consumer interval' if its consumer graph has this property, and 'resource interval' if its resource graph has this property. It is necessary now to look rather more closely at the various types of web, community, source and sink, as defined by Cohen. Any sink web is a sub-web of some community web. If a sink web is not consumer interval, then the addition of new consumers to make up the community web cannot change this property of the sub-web, and so the community web is not consumer interval. On the other hand if we supplement a source web by adding new sources to make up the community web, some components of the original source web will acquire new prey. Consequently there can be new cases of sharing a prey. So we can draw no conclusion about the consumer interval property of the community web. Reciprocally we can draw no conclusion about the resource interval property of the community web from the fact that a sink web is not resource interval. But if a source web is not resource interval, the addition of new sources to make up the community web cannot change this property of the sub-web, and so the community web is not resource interval.

Cohen presents two kinds of result bearing on the consumer interval property in community webs. There is a strong result, that seven out of nine single-habitat community webs are consumer interval. There is also a weak result, that thirteen out of fourteen single-habitat are consumer interval—which is consistent with the community webs from which they are drawn being consumer interval. I find that the same seven community webs are resource interval, and the other two are not. I find that all the nine available source webs (Askew 1971, Birkeland 1974, Force 1974, Hurlbert *et al.* 1972, Milne and Dunnett 1972, Richards 1926, Root 1975, Tilly 1968) are resource interval, which is consistent with the community webs from which they are drawn being resource interval.

Sugihara (1982) had access to a more extensive, and so far unpublished, compilation of foodweb data by F. Briand. He finds that only ten out of 73 community webs are not consumer interval. He points out that none of the community graphs has non-rigid circuits, the interval property being broken in all these ten cases by the occurrence of a star sub-graph.

Both Cohen and Sugihara find that the empirical foodwebs are significantly more likely to be consumer interval than are randomly constructed webs with similar numbers of components. They adopt very different viewpoints in discussing why this should be the case.

4.4 INTERPRETATIONS OF THE INTERVAL PROPERTY

Cohen (1978) takes literally the interpretation of an interval graph in terms of the overlap of two regions in an abstract resource space. The property does not compel one to use a space of one dimension, but allows one to do so if there is some other reason for this choice. Cohen considers three possible interpretations of restricted dimensionality of a resource space. One is the suggestion that, in his words,

> 'organisms may have more degrees of freedom in their physiological capacities to exist under varied circumstances than the biotic, especially trophic, interactions with other kinds of organism in the community permit them to enjoy.'

This merely seems to restate the assumption that low dimension is appropriate. No reason is given why this restriction by the other organisms should not merely reduce the volume occupied without a change of dimension. The second and third relate to possible connections of low dimension with stability, either in the sense, to be discussed in Chapters 6 and 8, of the sign stability of an equilibrium point, or in a sense connected with recurrence in random walks. Neither of these is pursued in any detail, and they do not seem particularly compelling.

Sugihara regards the rigid circuit property and the absence of star sub-graphs as significant independently of the interval aspect. He counts all cases of shared prey separately, so that the consumer graphs can be multi-edged. As an alternative realization of these competitive features he uses a simplicial complex (see Appendix 1). In this the resource niche of a species is a polyhedron with vertices associated with the prey species. These polyhedra can be in contact at a vertex (one prey shared) along an edge (between two vertices corresponding to two shared prey), and so on. The central question in the topology of clusters of polyhedra formed in this way is whether they are compact or contain any holes. Sugihara finds that holes are very rare, and relates this to the biological feature that species are added to existing communities in a conservative manner. (Two examples of simplicial complexes formed from three triangles, one with and one without a hole, are shown in Appendix 1.)

References

Askew, R. R. (1971). *Parasitic Insects.* Heinemann, London.

Benzer, S. (1959). On the topology of the genetic fine structure, *Proc. Nat. Acad. Sci. U.S.A.* **45**, 1607–1620.

Birkeland, C. (1974). Interactions between a sea pen and seven of its predators, *Ecol. Monogr.* **44**, 211–232.

Cohen, J. E. (1978). *Food Webs and Niche Space.* Princeton University Press, Princeton, N.J.

Force, D. C. (1974). Ecology of insect host–parasitoid communities, *Science* **184**, 624–632.

Fulkerson, D. R., and Gross, O. A. (1965). Incidence matrices and interval graphs, *Pac. J. Math.* **15**, 835–856.

Gilmore, P. C., and Hoffman, A. J. (1964). A characterisation of comparability graphs and of interval graphs, *Can. J. Math.* **16**, 539–548.

Hurlbert, S. H., Mulla, M. S., and Wilson, H. R. (1972). Effects of an organophosphide insecticide on the phytoplankton, zooplankton and insect populations of freshwater ponds, *Ecol. Monogr.* **42**, 269–299.

Lekkerkerker, C. G., and Boland, J. C. (1962). Representation of a finite graph by a set of intervals on the real line, *Fund. Math. Akad. Nauk. Polska* **51**, 45–64.

Milne, H., and Dunnett, G. M. (1972). Standing crop, productivity and trophic relationships in the fauna of the Ythan estuary, in *The Estuarine Environment* (edited by R. S. K. Barnes and J. Green), Applied Science Publications, Edinburgh, pp. 86–103.

Richards, O. W. (1926). Studies in the ecology of English lakes III, *J. Ecol.* **14**, 244–281.

Root, R. B. (1975). Some consequences of ecosystem texture, in *Ecosystem Analysis and Prediction* (edited by S. A. Levin), S.I.A.M. Philadelphia, pp. 83–92.

Sugihara, G. (1982). Niche hierarchy: structure, organisation and assembly in natural communities, Princeton University thesis.

Tilly, L. J. (1968). The structure and dynamics of Core Spring, *Ecol. Monogr.* **38**, 169–197.

Chapter 5

Complexity of trees and networks

Complexity is a multi-dimensional concept and there is unlikely to be a single satisfactory measure of this aspect of the structure of a system. In analysing a system one can choose various mathematical frameworks, within each of which one may assess the complexity of the system. These assessments may be very different. Even within one particular mathematical framework, although appropriate measures of complexity may readily be made, there is unlikely to be an obvious best measure. Casti (1979) surveys a number of treatments of complexity, appropriate, for example, when the mathematical framework involves algorithms, graphs, simplicial complexes (Appendix 1), linear or non-linear dynamical systems, and so on. I shall only discuss a few of these, with a bias towards those related to graphs.

The aspect of complexity which is most often studied is that related to the generation of a sequence of digits or symbols by an algorithm. Let $[m_1]$ and $[m_2]$ be two sequences of digits generated by algorithms which can themselves be coded by the sequence $]n_1[$ and $]n_2[$ of lengths L_1 and L_2. Then the first sequence is said to be the more complex of the two if $L_1 > L_2$. In this convention repetitiveness, which implies predictability, lowers the complexity of a sequence. For example, however large the integer p, the sequence

$$0101010101 \ldots p \text{ times}$$

can be coded as

$$01 : p$$

A random sequence $[m]$ is one for which the algorithm $]n[$ is about as long as the sequence itself. So one arrives at the rather disappointing conclusion that complexity is to be identified with randomness.

Shapes in two dimensions can be described by strings of symbols. Papentin (1980a) gives an instructive example, applying this method to compare the complexities of the outlines of a variety of polygonal figures representing snowflakes.

Photographs of snowflakes were projected onto transparent paper, outlined and transferred to a hexagonal grid of lines intersecting at 60° and 120°, supplemented when necessary by lines intersecting at 30° or 90°. The diameters were set equal to 108 grid units. The lengths of sides were recorded in grid units (to the nearest higher integer) and the turning angles coded as follows

$$60° = \alpha,\ 90° = \beta,\ 120° = \gamma,\ 150° = \delta,\ 210° = \delta',\ 240° = \gamma'$$
$$270° = \beta',\ 300° = \alpha'$$

Stepping round the outline gives, for the regular hexagon

$$54\gamma54\gamma54\gamma54\gamma54\gamma54\gamma \tag{5.1}$$

and, for the more elaborate outline in Fig. 5.1

$$\begin{gathered}18\gamma18\alpha'18\gamma18\gamma18\gamma18\alpha'18\gamma18\gamma18\gamma18\alpha'18\gamma18\gamma \\ 18\gamma18\alpha'18\gamma18\gamma18\gamma18\alpha'18\gamma18\gamma18\gamma18\alpha'18\gamma18\gamma\end{gathered} \tag{5.2}$$

On using the code $S = 54\gamma$, $S_1 = 18\gamma18$, and $S_2 = S_1\alpha'S_1\gamma$, these can be reduced to

$$6S:54\gamma \quad \text{and} \quad 6S_2:S_1\alpha'S_1\gamma::18\gamma18$$

When these descriptions have been made as short as possible the lengths of the shortened strings are compared to compare complexities of outlines. Repetition, associated with the typical hexagonal symmetry of the snowflakes, shortens strings. Thus an irregular outline requiring n symbols, with no possibility of shortening, can have similar complexity to an outline with six-fold symmetry requiring, before shortening, about $6n$ symbols.

It can be argued that since the 'typical' snowflake has hexagonal symmetry it may be more to the point not to reduce the string for repetitions or to use a measure (such as the ratio of the perimeter length estimated with a fine grid to the perimeter length measured with a coarse grid), making the more highly indented outlines more complex. The point is that some understanding of characteristic regularities of the systems under consideration should be built into the choice of a

Fig. 5.1 A snowflake outline with hexagonal symmetry

measure of complexity. Elsewhere Papentin (1980b) indicates that he is aware of this problem, and proposes that one should write complexity as the sum of two terms, one associated with the regularities in the structure, which may obey more or less complex rules, and the other with random departures from these regularities. For example, in dealing with foodwebs, the length of the longest chain of energy transfers is an aspect of complexity associated with the regularity: this is a trophic structure. The trophic impurity defined in Section 3.1 is an aspect of complexity associated with departures from this regularity. There are two difficulties here, the fundamental one of recognizing the nature of the essential regularities of the system, and the secondary one of how to weight the two contributions to the complexity.

5.1 TREES

The complexity of trees can be studied by the method employed by Papentin (1980b). A rooted tree can be described by a string of symbols. First label each edge as i (interior) or e (exterior). The **exterior** edges are those contiguous to the leaves. Starting at the root go clockwise round the tree, recording the labels as the edges are reached but ignoring any edge already traversed. The tree in Fig. 5.2**A** is constructed in a random manner. The one in **B** has a highly repetitive structure; for example, if cut at the node N it has two identical parts. Each of these trees is labelled by a string of 31 symbols, respectively

iiiieieeieeiiieiiieeeieeieieeee

and (5.3)

iiiieeieeiieeieeiiieeieeiieeiee.

The first of these strings can be slightly shortened, to 26 symbols, by noting the five-fold repetition of the string iee and using the convention that :fiee means ‘f stands for iee’. This string becomes

iiiieffiiieiifefiefee :fiee (5.4)

But the second string can be substantially reduced, to eighteen symbols, by a similar convention. It becomes

ihh : higg : giff :fiee. (5.5)

Trees without strict geometrical regularity can still have statistical regularity, as emphasized in Part III. Complexity of such trees is briefly considered in Section 12.3.

5.2 NETWORKS

Before going into more detail concerning the complexity of foodweb digraphs, it is pertinent to ask what graph theorists understand by the term complexity. Brooks *et al.* (1940) introduced the term **complexity** to stand for the number of spanning

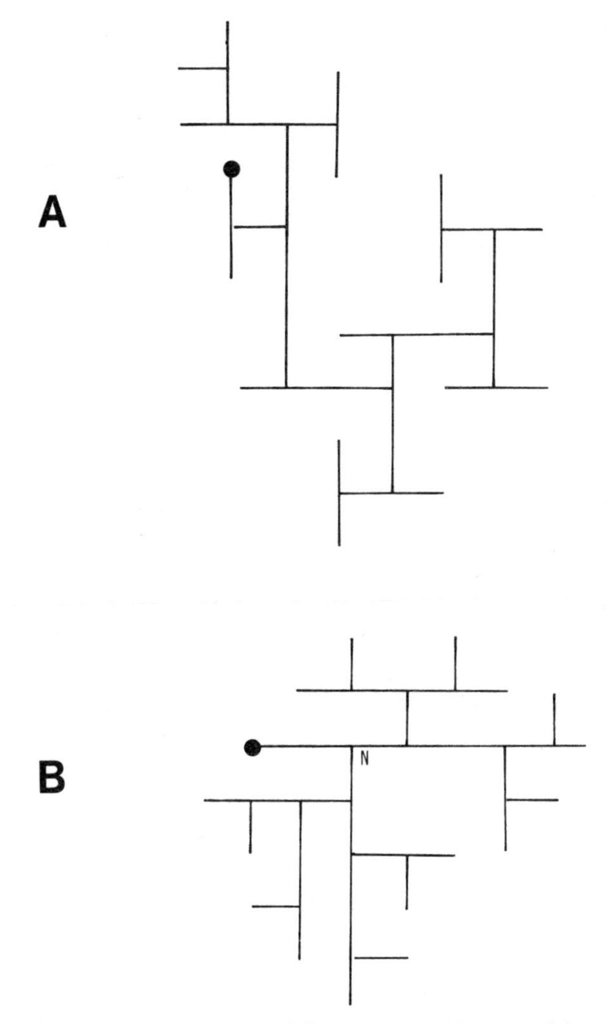

Fig. 5.2 A random tree (**A**) and a repetitive tree (**B**), each with 31 edges. The root is marked with a spot in each case. If the tree **B** is cut at the node N it falls into two identical parts

trees in a graph. A **spanning tree** T of a graph G is a tree in which the vertices are all the vertices of G, and all the edges are among the edges of G. If G has many cycles there will be many ways of finding a T. If G has many cycles counting the various T sounds a highly complicated task, but an algorithm is available which illustrates the advantages of working with the adjacency matrix $\boldsymbol{A}$. Biggs (1974) gives the complexity $\boldsymbol{K}$ of a graph with V vertices as

$$\boldsymbol{K} = V^{-2} \det[\boldsymbol{J} + \boldsymbol{D} - \boldsymbol{A}] \tag{5.6}$$

where

$J = V \times V$ matrix with all elements $= 1$

$A =$ the adjacency matrix

$D =$ a diagonal matrix with $d_{ii} =$ the number of edges ik, known as the **valency** of i

For a fully connected graph, with V vertices and $\frac{1}{2}V(V-1)$ edges, the value of K is

$$V^{-2} \det(D + I) = V^{V-2} \tag{5.7}$$

where I is the diagonal unit matrix.

There are some papers on the complexity of graphs from the point of view of information theory (Mowshowitz 1968). However, this can only deal with Papentin's second type of complexity, associated with departures from regular structures. For example, our foodweb example of Chapter 3, which when analysed as a 3-level pure trophic structure had five, six and two components in the three levels. In a trivial information theory sense this is more complex than a 12-component 3-level web with four components in each level.

One aspect of complexity in foodwebs (see Gallopin (1972), MacDonald (1979)) is the extent to which the number of food chains, leading from the lowest trophic level to the highest, measures up to the maximum number possible for the given trophic structure. Consider a pure trophic digraph with the nodes divided into sets $I_1, I_2, \ldots, I_P$ and arcs only present from I_r to I_{r+1}. Let the numbers of nodes in each level be $N_1, N_2, \ldots, N_P$. The maximum possible number of chains, attained when every node in I_{r+1} receives an arc from each node in I_r, is

$$K = \prod_{i=1}^{P} N_i \tag{5.8}$$

A measure of chain redundancy, as an aspect of complexity, would be the actual number of chains divided by K.

Foodwebs are impure trophic structures, and this measure should be amended to allow for this fact. Gallopin (1972) defines a **trophic partition** of a digraph, in which nodes are divided into sets $L_1, L_2, \ldots, L_Q$ such that arcs occur from members of L_r to members of $L_{r+1}, L_{r+2}, \ldots$ members of L_Q. Let the numbers of nodes in these sets be $M_1, \ldots, M_Q$. The maximum number of chains is

$$K' = M_1 M_Q \left[1 + \sum_{j=2}^{Q-1} M_j + \text{sum of products } M_j M_k + \cdots + \prod_{j=2}^{Q-1} M_j \right] \tag{5.9}$$

In large webs the assignment to sets L_j can be ambiguous when one is equally far from L_1 and L_Q. I have calculated (MacDonald 1979) the ratio of the number of chains to K' for the webs compiled by Cohen (1978). I made assignments to L_j so as to reduce the range of variation of this ratio, and still found that the largest webs could have values three orders of magnitude below those for small webs.

In the following chapter I shall describe the theoretical context in which questions about the complexity of foodwebs have usually been posed. I shall give reasons for favouring, in that context, a simple and unambiguous one, the ratio A/N of the number of arcs to the number of nodes.

References

Biggs, N. L. (1974). *Algebraic Graph Theory*, Cambridge University Press.

Brooks, R. L., Smith, C. A. B., Stone, A. H., and Tutte, W. T. (1940). The dissection of rectangles into squares, *Duke Math. J.* **7**, 312–340.

Casti, J. (1979). *Connectivity, Complexity and Catastrophe in Large Scale Systems*, Wiley, New York.

Cohen, J. E. (1978). *Foodwebs and Niche Space*, Princeton University Press, Princeton, N.J.

Gallopin, G. C. (1972). Structural properties of foodwebs, in *Systems Analysis and Simulation in Ecology* Vol. II (edited by B. C. Patten), Academic Press, pp. 241–283.

MacDonald, N. (1979). Simple aspects of foodweb complexity, *J. theor. Biol.* **80**, 577–588.

Mowshowitz, A. (1968). Entropy and the complexity of graphs, I and IV, *Bull. Math. Biophysics* **30**, 175–204 and 535–546.

Papentin, F. (1980a). Complexity of snowflakes, *Naturwiss* **67**, 174–177.

Papentin, F. (1980b). On order and complexity I, *J. theor. Biol.* **87**, 421–456.

Part II

Stability of equilibrium states and of periodic behaviour

'But in spite of all temptations,
To belong to other nations,
He remains an Englishman!'

Sir W. S. Gilbert, *H.M.S. Pinafore*

Chapter 6

Stability and complexity of model foodwebs

There has been a long debate among ecologists on the question whether the complexity of a natural community assists it to be stable. This debate was focused in particular on the concepts introduced by MacArthur (1955). He defined complexity as the amount of choice of the energy's path going through the foodweb, and stability in this way:

> 'Suppose for some reason that one species has an abnormal abundance, then we shall say that the community is unstable if the other species change markedly in abundance as a result of the first. The less effect this abnormal abundance has on the other species the more stable the community.'

The state of the debate twenty years after MacArthur's paper was reviewed by Goodman (1975).

Mathematical work on this question requires a choice of an appropriate definition of complexity and a choice of an appropriate definition of stability. A rather straightforward definition of complexity is suggested by an important paper by May (1972). In this paper, as in many of the subsequent contributions, stability is interpreted as the local stability of an equilibrium point. When an arbitrary small disturbance is made in one or more of the populations, away from the equilibrium value or values, the populations return towards their equilibrium values. Another popular definition of stability in this context, which Pimm (1979) for instance claims to be the nearest practical equivalent to MacArthur's original formulation, is species deletion stability. A set of N species populations with equilibrium values is disturbed by completely removing the population of one species. If the system is stable against species deletion then, whatever the choice of deleted species, a new stable equilibrium of the remaining $N-1$ species is attained.

Again the addition of a new species affords another definition of stability. A new stable equilibrium of $N+1$ species can be attained, or one or more species may be eliminated. Natural experiments, due to the inadvertent introduction of a new predator species, occur from time to time. One such has been studied by Zaret (1982) in the freshwater communities in a South American lake and associated rivers. The lake community is the more diverse. The river communities lost no species when the predator was introduced. The lake lost thirteen out of seventeen species of fish.

Let us look in particular at the deletion or addition of a top predator, that is one not itself subject to predation. If the population of this predator has no significant effect on the populations of its prey, the interactions of this predator and its prey are said to be donor controlled. Then the foodweb must be stable under either the addition or the deletion of the top predator. Donor control can apply, for example, if predation is confined to dead, sick, and injured prey. Otherwise it is an artificial assumption, and in general I shall assume two-way prey–predator interactions, as in models of the Lotka–Volterra type.

6.1 TWO TYPES OF MODEL

When local stability is studied there is a further choice, between what I shall call Type 1 and Type 2 models. A model of Type 1 starts from an assumed set of rate equations for the populations. Dealing for the moment with overlapping generations and a continuous time (ordinary differential equation) formulation, these rate equations are

$$\frac{dx_i}{dt} = f_i(x_1, \ldots, x_N), \quad i = 1, \ldots, N \tag{6.1}$$

The equilibrium populations x_i^0 are calculated from

$$\left.\frac{dx_i}{dt}\right|_{x_1=x_1^0,\ldots,x_N=x_N^0} = f_i(x_1^0, \ldots, x_N^0) = 0 \tag{6.2}$$

$$i = 1, \ldots, N$$

Local deviations from the equilibrium are described in terms of new variables

$$X_i = x_i - x_i^0, \quad i = 1, \ldots, N$$

To a linear approximation these new variables satisfy rate equations

$$\frac{dX_i}{dt} = \sum_j a_{ij} X_j, \quad i = 1, \ldots, N \tag{6.3}$$

where the constant coefficients a_{ij} are the partial derivatives of the functions f_i with respect to the variable x_j, evaluated at the equilibrium point

$$a_{ij} = \left.\frac{\partial f_i}{\partial x_j}\right|_{x_1=x_1^0,\ldots,x_N=x_N^0} \tag{6.4}$$

In matrix form the rate equations (6.3) are

$$\frac{d\boldsymbol{X}}{dt} = A\boldsymbol{X} \tag{6.5}$$

and the matrix A with elements a_{ij} is known as the Jacobian matrix. Stability is tested by the requirement that all roots λ of the equation

$$\det(A - \lambda I) = 0 \tag{6.6}$$

the secular equation, must have negative real parts. In equation (6.6) I stands for the $N \times N$ matrix with diagonal elements 1 and other elements 0, and 'det' stands for the determinant. This quantity is an Nth-order polynomial in λ; its properties will be discussed extensively in the next chapter.

Most commonly, in models of Type 1, a particular choice of the functions f_i is made, so that the rate equations are of the Lotka–Volterra type

$$\frac{dx_i}{dt} = x_i\left[r_i + \sum_j b_{ij}x_j\right], \quad i = 1, \ldots, N \tag{6.7}$$

The appearance of x_i as a factor is a simple way of ensuring that if one starts with positive x_i these quantities stay positive. The expression in square brackets can be thought of as the sum of birth rate and death rate per individual of species i. The particular form chosen has the considerable advantage that the calculation of equilibrium values x_i^0 is a linear one. In matrix notation

$$\boldsymbol{x}^0 = -\boldsymbol{B}^{-1}\boldsymbol{r} \tag{6.8}$$

and there is a unique solution, with no x_i^0 zero, so long as $\boldsymbol{B}$, the matrix with elements b_{ij}, is non-singular. The matrices $\boldsymbol{A}$ and $\boldsymbol{B}$ are related by

$$a_{ij} = x_i^0 b_{ij} \tag{6.9}$$

It should be noted that equations of the form (6.7) also have equilibrium points with one or more of the variables taking value zero. These, of course, are relevant to the question of stability under species deletion or species addition.

At present we are concerned with local stability. In a Type 1 model the first step is to find the equilibrium point, the next to calculate the a_{ij}, and the final step is to analyse the secular equation. In a Type 2 model the first step is left out, and the starting point is to express assumptions about interactions by means of assumptions about the values of the coefficients a_{ij}. Type 1 models allow, or force, one to be rather more explicit about these assumptions, and they are of course necessary if stability under species deletion or addition is to be examined. One further important point must be made about models of Type 1. Populations are always positive, and as mentioned for the Lotka–Volterra form, this can be ensured in numerical integration of the rate equations by suitable choice of the f_i. But the calculation of equilibrium values, for example by using equation (6.8), carries no guarantee that these will all be positive. So one must discard any equilibrium point with negative population. It can readily be shown (Goh and Jennings 1977) that only one in 2^N cases is retained for Lotka–Volterra models with random signs for the r_i and b_{ij}. Those retained are usually called, following Roberts (1974), feasible equilibrium points.

6.2 STABILITY AND CONNECTEDNESS

Gardner and Ashby (1970), in connection with a general investigation of connectedness and stability in complex systems, not specifically directed at ecological communities, presented numerical results for a Type 2 model with N up to ten. They took a_{ii} in the range –0.1 to –1.0, so that all components are self-stabilized in isolation. A certain fraction C of the a_{ij} $(i \neq j)$ were taken to be non-zero and given values distributed evenly between –1.0 and 1.0. For sets of ten equations, the probability of local stability fell from near 1 for small C to near 0 for large C, the fall being sharply concentrated around $C = 0.13$.

May (1972) used a Type 2 model and obtained analytic asymptotic results for large N. He set all $a_{ii} = -1$, so that the uncoupled equations each have real negative root –1. He took a fraction C of the a_{ij}, $(i \neq j)$ to be non-zero, and gave them equal probability of positive or negative values, with root mean square value a. He used a result on the distribution of the characteristic roots of a symmetric random matrix (Wigner's semicircle law) to show that the largest real root of the coupled system lies near

$$-1 + Na^2 C$$

with a high probability, for large N. So the transition from expected stability to expected instability is sharp, and the critical value of C is about

$$C^* = 1/Na^2 \qquad (6.10)$$

Comparing this with Gardner and Ashby's value, with $N = 10$ and $a^2 = 1/3$ and mean uncoupled root –0.55, one has 0.16 instead of 0.13. May only sketched a proof of this random matrix result; details are given by Hastings (1982).

If we choose to define our measure of complexity as NC, we can state May's result as: complexity greater than a^{-2} yields (almost certainly for large N) instability. For the foodwebs discussed in earlier chapters we of course have no way of assessing a, but this result suggests that a meaningful, as well as a simple, complexity measure will be NC. For a digraph with N nodes and A arcs, this is

$$N[2A/N(N-1)] = 2A/(N-1) \qquad (6.11)$$

For fairly large webs this is not much different from $2A/N$. A number of authors (MacDonald 1979, Pimm 1980, Yodzis 1980) have pointed out that the values of this quantity cluster around 4 (± 1) in the webs of Cohen's compilation (Cohen 1978) supplemented by some more webs, mainly source webs which Cohen ignores. This regularity has also been pointed out (Rejmanek and Stary 1979) in a collection of about thirty webs drawn from data on plant–aphid–parasitoid communities. Thus over quite a wide range of N, from ten to 100, empirical webs have a rather narrow range of values of this particular measure of complexity.

Roberts (1974) presented a Type 1 model, of the Lotka–Volterra type with positive r_i and b_{ij} equally likely to take values $\pm z$, and emphasized the necessity for removing all but the feasible equilibrium points. He found that in feasible cases stability did not fall off with complexity, and considered that this invalidated May's results. However, Gilpin (1975) pointed out that Roberts' choice of parameters was rather inappropriate. In particular, the positive r_i imply that predator populations grow in the absence of prey. He found that a Lotka–Volterra model with random signs gave results consistent with those of May after restriction to the feasible cases. This was confirmed by Goh and Jennings (1977) using larger sets of random models.

6.3 SELF-STABILITY, OMNIVORY, COMPARTMENTALIZATION

Apart from the random matrix result of May (1972), there is not much that can be said without detailed numerical work. However, one can, with due caution as stressed by May (1973), point to the requirements for a linearized set of ordinary differential equations to be sign stable. This means that stability conditions are specified in terms of certain a_{ij} being zero and others having fixed sign, the magnitude of these not being known. These matters will be looked at in more detail in Section 8.1, and here only the relevant results are quoted. Sign stability follows if

$$\begin{aligned}
&a_{ii} \leqslant 0 \text{ for all } i, \text{ and for at least one } i,\ a_{ii} < 0 \\
&a_{ij}a_{ji} \leqslant 0 \text{ for all } i, j \neq i \\
&\text{all higher products such as } a_{ij}a_{jk}a_{ki}, \text{ where } i \neq j,\ i \neq k \text{ and } j \neq k, \\
&\qquad \text{are identically zero} \\
&\det A \neq 0
\end{aligned} \tag{6.12}$$

So a model foodweb, with self-interactions either zero or stabilizing, and with no omnivory, in the sense that 'i eats j', 'j eats k' and 'i eats k' never occur together, is guaranteed to have local stability.

These results have the merit of drawing attention to two topics that have turned out to be significant in more recent studies of foodweb stability, the importance of self-stabilizing components and the possibility that omnivory is destabilizing. Another matter that was seen to require investigation, as a consequence of May's (1972) paper, is compartmentalization. May pointed out that in the Gardner and Ashby (1970) numerical Type 2 results, a community with twelve species with $C = 0.15$ has essentially zero probability of local stability. However, if the community is compartmentalized into three blocks of four species each, with no inter-block interactions and so a value of C, within each block, of 0.45, 35% of cases are locally stable.

All recent work in this field starts from the premise that, whether Type 1 or Type 2 models are used, these should not have random patterns of interaction but should include biologically realistic general features. For example, one can exclude loops such as 'i eats j, j eats k, k eats i'. One can ensure a bias in the relative magnitudes of the equilibrium populations, predators being less abundant than their prey. For example, Pimm (1979) presents a Type 2 model in which for any pair i—predator j—prey, $a_{ij} \ll |a_{ji}|$. On the assumption of an underlying Lotka–Volterra Type 1 model, this is interpreted as the consequence of

$$b_{ij} \doteqdot |b_{ji}|, \quad x_i^0 \ll x_j^0$$

In models of this kind Pimm finds that locally stable cases differ from the average by having greater trophic purity (fewer omnivores) and more cross-links between food chains (less compartmentalization) for a given fraction C of non-zero interactions.

Pimm (1980) finds that real foodwebs differ from randomly generated ones, with the same N and C, by having fewer trophic levels and greater trophic purity. Pimm and Lawton (1980) find it difficult to get good evidence on compartmentalization in real foodwebs, but conclude that it does not seem to be a general feature. Pimm (1979, 1980) finds general consistency between the requirements for local stability in Type 2 models, and the structure of real webs, and no such consistency when the requirements for stability are those for species deletion stability for Type 1 models.

There is some disagreement about the extent to which self-stabilizing species occur at any trophic level other than the lowest. Pimm and Lawton (1978) argue that self-stabilization can only be a feature of species that have quite sophisticated social patterns such as hierarchy or territoriality, and that crop a rather low proportion of the prey populations. The question matters to their conclusions, and in particular regarding the length of food chains. In their models stability depends on the self-stability at the lowest level, and this stabilizing effect is more attenuated the longer the chain. Ways in which the characteristics of predator or parasitoid populations can exercise a stabilizing effect in a food chain are the subject of some controversy (Abrams and Allison 1982).

In all this recent work there has been a movement away from the formulation of general questions about stability and complexity towards the detailed investigation of which specific features of the structure of ecological communities, in particular of foodwebs, may be correlated with stability. Because of the bias already mentioned towards analysis of systems in terms of distinct parts and their interactions, it is not surprising that general treatments of the stability of complex systems (Siljak 1978, Michel and Miller 1977) deal entirely with the case of stable subsystems connected by potentially destabilizing interactions. While these authors go well beyond the narrow definition of stability adopted here, as local stability of an equilibrium point, they use throughout the assumption of compartmentalization. Roberts and Tregonning (1980) offer a suggestion of an alternative style of stable structure with nested subsystems. However, their work is marred by what seems an excessive reliance on the identification of (stable or unstable) feasible equilibrium points, as equivalent to the identification of stable solutions such as limit cycles. The remarks of Sugihara (1982) about conservative addition of species to communities are clearly relevant to studies of species deletion and species addition.

An excellent survey of foodweb structure and stability is given in the recent book by Pimm (1982).

References

Abrams, P. A., and Allison, D. (1982). Complexity, stability and functional response, *Amer. Nat.* **119**, 240–249.

Cohen, J. E. (1978). *Foodwebs and Niche Space*, Princeton University Press, Princeton, N.J.

Gardner, M. R., and Ashby, W. R. (1970). Connectance of large dynamic (cybernetic) systems: critical value for stability, *Nature* **228**, 784.

Gilpin, M. E. (1975). Stability of feasible prey–predator systems, *Nature* **254**, 137–139.

Goh, B. S., and Jennings, L. S. (1977). Feasibility and stability in randomly assembled Lotka–Volterra models, *Ecological Modelling* **3**, 63–71.

Goodman, D. (1975). The theory of diversity–stability relationships in ecology, *Quart. Rev. Biol.* **50**, 237–266.

Hastings, H. M. (1982). The May–Wigner stability theorem, *J. theor. Biol.* **97**, 155–166.

MacArthur, R. H. (1955). Fluctuations in animal populations and a measure of community stability, *Ecology* **36**, 533–536.

MacDonald, N. (1979). Simple aspects of foodweb complexity, *J. theor. Biol.* **80**, 577–588.

May, R. M. (1972). Will a large complex system be stable? *Nature* **238**, 413–414.

May, R. M. (1973). *Stability and Complexity in Model Ecosystems*, Princeton University Press, Princeton, N.J.

Michel, A. N., and Miller, R. K. (1977). *Quantitative Analysis of Large Scale Dynamical Systems*, Academic Press, New York.

Pimm, S. L. (1979). Complexity and stability—another look at MacArthur's original hypothesis, *Oikos* **33**, 351–357.

Pimm, S. L. (1980). Properties of foodwebs, *Ecology* **61**, 219–225.

Pimm, S. L. (1982). *Food Webs*, Chapman and Hall, London.

Pimm, S. L., and Lawton, J. H. (1978). Reply: population dynamics and the length of food chains, *Nature* **272**, 190.

Pimm, S. L., and Lawton, J. H. (1980). Are food webs divided into compartments? *J. An. Ecol.* **49**, 879–897.

Rejmanek, H., and Stary, P. (1979). Connectance in real biotic communities and critical value for the stability of model ecosystems, *Nature* **280**, 311–313.

Roberts, A. (1974). The stability of a feasible random ecosystem, *Nature* **251**, 607–608.

Roberts, A., and Tregonning, K. (1980). The robustness of natural systems, *Nature* **288**, 265–266.

Siljak, D. D. (1978). *Large Scale Systems: Stability and Structure*, North-Holland, Amsterdam.

Sugihara, G. (1982). Niche hierarchy: structure, organisation and assembly in natural communities, Princeton University thesis.

Yodzis, P. (1980). The connectance of real ecosystems, *Nature* **284**, 544–545.

Zaret, T. J. (1982). The stability–diversity controversy: a test of hypothesis, *Ecology* **63**, 721–731.

Chapter 7

Graphical aspects of local stability theory: cycle analysis

The local stability of an equilibrium point of a system of ordinary differential equations (o.d.e.) or of difference equations is discussed in many texts. Less frequently discussed is the possibility of understanding local stability in terms of the structure of the system of equations. In biological applications, in which the number of coupled equations may be quite small, but the parameter values are rarely precisely known, this is particularly useful. Frequently one wishes to survey a parameter space and judge what changes in stability are likely. In recent years certain authors (Levins 1974, 1975, Roberts 1976, Roberts and Brown 1975) have emphasized how to relate the familiar criteria for stability to cycles in the network of interactions among the components of the system.

Another aspect of local stability which is not always sufficiently emphasized is the qualitative difference between the stability requirements for o.d.e. and for apparently analogous difference equations. The cycle analysis of Levins has sometimes been presented as if it had a direct bearing on difference equations, although results which are necessary and sufficient for local stability of an equilibrium point in the o.d.e. case are likely to give a rather weak necessary condition in the difference equation case.

Let us start from the system of equations linearized around an equilibrium point. As in Chapter 6, for o.d.e. this is given in terms of the Jacobian matrix A

$$\frac{d\boldsymbol{X}}{dt} = A\boldsymbol{X} \tag{7.1}$$

For difference equations

$$x_{t+1}^{i} = f_i(x_t^1, \ldots, x_t^N) \tag{7.2}$$

with the equilibrium point defined by

$$x_t^{i0} = f_i(x_t^{10}, \ldots, x_t^{N0}) \tag{7.3}$$

the linearized equations for the new variables $X_i^t = x_t^i - x_t^{i0}$ are

$$X_{t+1} = A\,X_t \tag{7.4}$$

The elements a_{ij} of the Jacobian matrix can be associated as weights with the arcs (j, i), going from j to i, of a digraph. The direction is specified in this way because a_{ij} quantifies the local dependence of the function f_i on the variable x_j, that is the influence of j on i. It is only in the linearized system that it is appropriate to use a weighted digraph description.

It is worth emphasizing that in general both arcs (i, j) and (j, i) are likely to be present. If a_{ij} is not zero, then in many models a_{ji} is also not zero. Taking an ecological community as an example, with the variables x_i interpreted as the populations of species i, predation requires a_{ij} and a_{ji} of opposite sign, and competition requires both a_{ij} and a_{ji} to be negative. So 2-cycles $a_{ij}a_{ji}$ are typically present, as well as 1-cycles (loops a_{ii}). Longer cycles, such as $a_{ij}a_{jk}a_{ki}$, will be present, for example, when there is a shared prey or a shared predator or a set of three mutual competitors. An exception to the prevalence of two-way interactions in this context is the phenomenon of amensalism, in which one species has a detrimental effect on the other without apparently a reciprocal influence of either a positive or negative kind. Another is donor-controlled predation.

Another quite different type of cycle structure is encountered in models of biochemical control processes (see, for example, Tyson and Othmer (1978)). In these models there is likely to be a chain of chemical species each influencing the rate of production of the next, while the final member of the chain inhibits or activates the production of the first. So for a single chain, and allowing for the decay of each member as a negative 1-cycle a_{ii}, the digraph for the local stability analysis of this model is as shown in Fig. 7.1. If any common members appear in two or more such chains, and this undoubtedly is the rule in cell biochemistry, then a complex structure of cycles is to be expected.

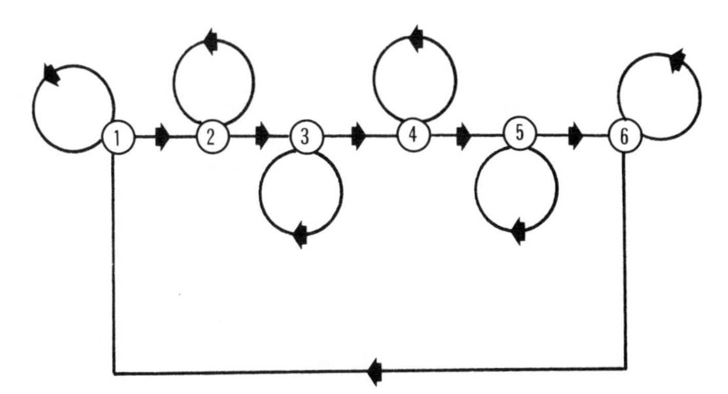

Fig. 7.1 The cycle structure of a Goodwin oscillator with six linear stages. There are six loops and one cycle of length six

7.1 THE SECULAR EQUATION

The central result of local stability theory employs the secular equation

$$P(\lambda) = \det(A - \lambda I) = \pm(\lambda^n + a_1\lambda^{n-1} + a_2\lambda^{n-2} + \cdots + a_{n-1}\lambda + a_n) = 0 \tag{7.5}$$

For o.d.e. local stability requires $\mathrm{Re}(\lambda) < 0$ for all roots λ. For difference equations local stability requires $|\lambda| < 1$ for all roots. From the definition of the determinant a number of useful results follow.

(1) No interaction a_{ij} appears in the secular equation unless the arc (j, i) forms part of a cycle beginning and ending at the same node. (The 1-cycle for a_{ii} is included among these.)

(2) If the digraph divides into two parts, linked only by arcs that do not form part of any cycle, then the secular determinant factors into two parts

$$\det(A - \lambda I) = \det(A_1 - \lambda I_1)\det(A_2 - \lambda I_2) \tag{7.6}$$

In this equation A_1 and A_2 are the matrices corresponding to the two separate digraphs, and I_1 and I_2 have the same dimension as A_1 and A_2 respectively. So there has to be a reciprocal interaction between two subsystems if the stability analysis for the whole system is not to reduce to the separate analyses for each of the sub-systems.

(3) No node label i appears more, or less, than twice in any product of the a_{ij} that occurs in any of the a_r.

(4) Each a_r, that is the coefficient of the $(n-r)$th power of λ, in the polynomial $P(\lambda)$, is a product of r factors a_{ij}, or a sum of such products. (Thus, for example, if all the a_{ij} are identical to a, $P(\lambda) = \lambda^n \times a$ polynomial in a/λ.)

(5) As a direct consequence of (3), where products of cycles occur in a_r these cycles are disjoint (have no common node).

(6) A general form can be set down for a_r. First, it is convenient to adopt the convention that the sign of the term of order n in λ, instead of being $(-1)^n$, is always positive. For a term in a_r having M disjoint cycles, define the parity as $(-1)^M$. Also, for any such term define the weight $W(L)$ as the product of the r factors a_{ij}. Sum all such weights having the same M, then multiply by the parity and sum over all distinct M that appear in a_r. This gives the central result of the cycle analysis of the secular equation

$$a_r = \sum_{M(r)} (-1)^M \sum_{L(M)} W(L) \tag{7.7}$$

Equation (7.7) is sufficiently important for it to be worth setting down in full in a moderately complicated case, $n = 4$. Here

$$\lambda^4 + a_1\lambda^3 + a_2\lambda^2 + a_3\lambda + a_4 = 0 \tag{7.8}$$

with

$$a_1 = -\sum_i a_{ii} \tag{7.9a}$$

(loops (1-cycles) only)

$$a_2 = -\sum_{i \neq j} [a_{ij}a_{ji} - a_{ii}a_{jj}] \tag{7.9b}$$

(2-cycles, and second-order effect of loops)

$$a_3 = -\sum_{\substack{j \neq i \\ j \neq k \\ i \neq k}} [a_{ij}a_{jk}a_{ki} - a_{ij}a_{ji}a_{kk} + a_{ii}a_{jj}a_{kk}] \tag{7.9c}$$

(3-cycles, second- and third-order effects)

$$a_4 = - \sum_{\substack{\text{all indices} \\ \text{unequal}}} [a_{ij}a_{jk}a_{ke}a_{ei} - a_{ij}a_{jk}a_{ki}a_{ee}$$

$$-a_{ij}a_{ji}a_{ke}a_{ek} + a_{ij}a_{ji}a_{kk}a_{ee}] - a_{11}a_{22}a_{33}a_{44} \tag{7.9d}$$

(4-cycles, two different types of second-order effect, third-order effects and a unique fourth-order term)

A diagrammatic mnemonic for this is set out in Fig. 7.2.

Interpreting the stability requirements on the coefficients a_r in terms of the structure of the interaction digraph is complicated, primarily because of the profusion of higher-order effects of short cycles. Particular simplifications come about if the digraph is a **rosette** (Pitts 1942; Roberts 1976) in which there are no disjoint cycles. In particular, this has no loops, except at the node which is common to all the cycles. Two examples are given in Fig. 7.3. Roberts in fact calls that in **A**, in which no node other than that common to all cycles is common to any pair of cycles, a rosette and that in **B** an advanced rosette, but I shall not retain this distinction. For these rosettes of order 4, the various coefficients in equations (7.9) lose all but their first terms, because of rule (5). So the coefficient a_r depends only on the cycles of length r.

Before setting down the standard results regarding stability in terms of the coefficients a_r, this is a convenient point to give the useful results for these coefficients in terms of the roots λ_i of equation (7.5)

$$\begin{aligned} a_1 &= -\text{sum of all roots } \lambda_i \\ a_2 &= +\text{sum of all products } \lambda_i\lambda_j \text{ of different roots} \\ a_3 &= -\text{sum of all products } \lambda_i\lambda_j\lambda_k \text{ of different roots} \end{aligned} \tag{7.10}$$

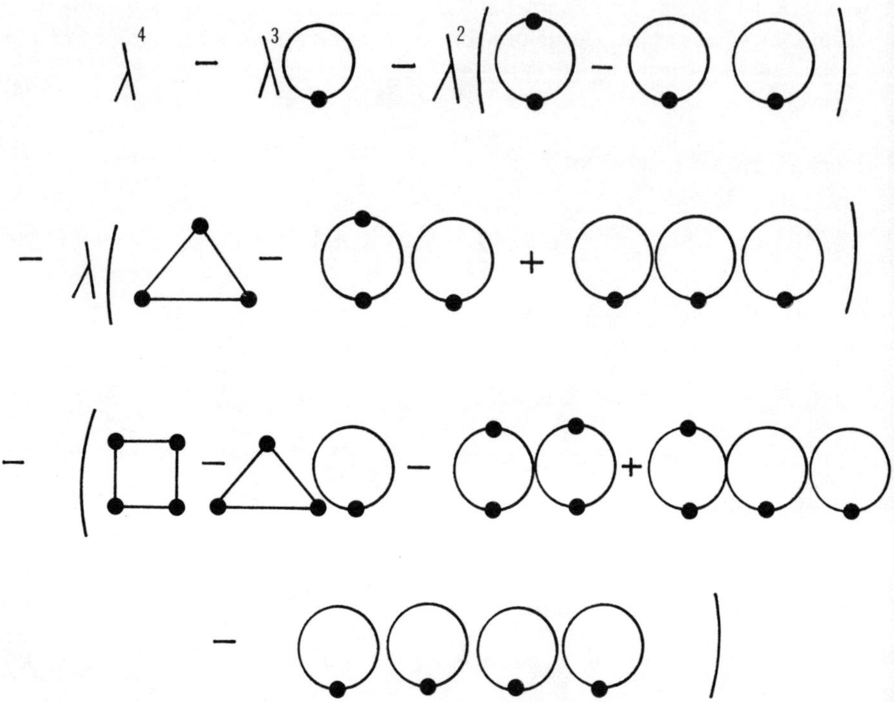

Fig. 7.2 A graphical mnemonic for the secular equation of a fourth-order set of coupled o.d.e.

and so on, and finally

$$a_n = \text{the product of all } n \text{ roots} \times (-1)^n$$

It is also worth reminding the reader at this point that roots are either real or complex, and that if complex they come in pairs which are complex conjugates. The real part of a root changing sign, which can signal a change in stability, can mean either a zero root or a pure imaginary pair of roots.

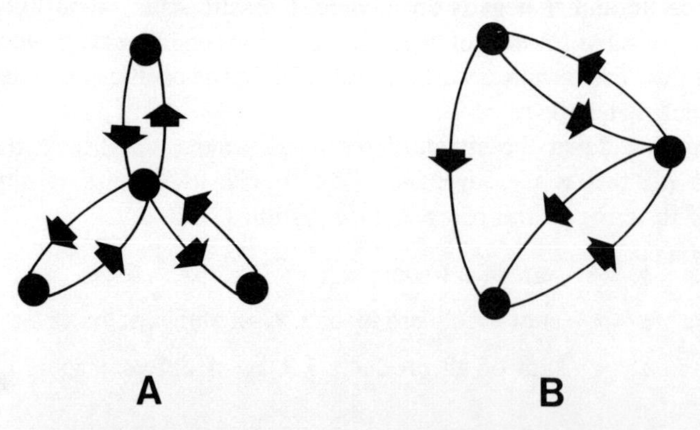

Fig. 7.3 A rosette (**A**) and an advanced rosette (**B**)

The local stability conditions for o.d.e. are set out very fully in Gantmacher (1959). First, a necessary condition for the real parts of all roots λ to be negative is that all coefficients a_r are positive. Second, given that all a_r are positive, a necessary and sufficient condition is that all the Routh–Hurwitz determinants

$$\Delta_{n-1}, \Delta_{n-3}, \ldots, \Delta_3 \text{ or } \Delta_2 > 0 \tag{7.11}$$

These determinants are defined by

$$\Delta_s = \begin{vmatrix} a_1 & a_3 & a_5 & \cdots & a_{2s-1} \\ 1 & a_2 & a_4 & \cdots & a_{2s-2} \\ 0 & a_1 & a_3 & \cdots & a_{2s-3} \\ 0 & 1 & a_2 & \cdots & s_{2s-4} \\ & & & & \\ & & & & a_s \end{vmatrix} \tag{7.12}$$

So

$$\Delta_2 = a_1 a_2 - a_3 \tag{7.13}$$

$$\Delta_3 = (a_1 a_2 - a_3)\, a_3 - a_1^2 a_4 \tag{7.14}$$

$$\Delta_4 = \begin{vmatrix} (a_1 a_2 - a_3) & (a_1 a_4 - a_5) & 0 \\ a_1 & a_3 & a_5 \\ 1 & a_2 & a_4 \end{vmatrix} \tag{7.15}$$

$$= (a_1 a_2 - a_3)(a_3 a_4 - a_2 a_5) - (a_1 a_4 - a_5)^2 \tag{7.16}$$

Higher ones get rather clumsy.

It is worth asking whether, without going through the proofs, one can see why these quantities matter for stability. The results in equation (7.10) indicate that the sign of a_n is important, since this coefficient is $(-1)^n \times$ the product of the real roots $\times$ a number of positive quantities (the moduli of the complex roots). So when all real roots are negative, which must mean either n, $n-2, \ldots$ negative factors, a_n is positive. When a single real root becomes positive, a_n becomes negative, but a_n does not change sign when the real part of a complex pair of roots changes sign. Let us look at the conditions for a pure imaginary root $+\omega$, for $n = 3$, 4, and 5. The polynomial $P(\lambda)$ can be written in the form

$$P(\lambda) = \lambda Q(\lambda) + R(\lambda) \tag{7.17}$$

in which the polynomials $Q(\lambda)$ and $R(\lambda)$ are even. For $n = 3$ the conditions for a root $+\omega$ are

$$Q(i\omega) = a_2 - \omega^2 = 0$$

$$R(i\omega) = a_3 - a_1 \omega^2 = 0$$

and these are consistent if Δ_2, as given by equation (7.13), is zero.

For $n = 4$ the conditions are

$$R(i\omega) = a_4 - a_2\omega^2 + \omega^4 = 0$$
$$Q(i\omega) = a_3 - a_1\omega^2 = 0$$

and these are consistent if

$$[a_3/a_1]^2 - a_2[a_3/a_1] + a_4 = 0$$

that is, if Δ_3, as given by equation (7.14), is zero. Finally, for $n = 5$ the conditions

$$Q(i\omega) = a_4 - a_2\omega^2 + \omega^4 = 0$$
$$R(i\omega) = a_5 - a_3\omega^2 + a_1\omega^4 = 0$$

These are consistent if $(a_1 a_2 - a_3) = 0$, and $(a_1 a_4 - a_5) = 0$, which from equation (7.15) give $\Delta_4 = 0$. So zero values of these Routh–Hurwitz determinants occur when the real part of a pair of complex roots is zero.

For rosettes there is a simple interpretation of the condition that the coefficients a_r are all positive. This implies that the magnitude of the resultant weight of any set of cycles of length r which each have negative weight must exceed the resultant weight of the cycles of the same length which have positive weight. The Routh–Hurwitz determinants display, when they are positive, a predominance of shorter cycles over longer ones, but not in a linearly additive manner.

One point in common between the conditions $a_r > 0$ and the conditions $\Delta_s > 0$ emerges if we take a set of equations for which all the a_{ij} are either zero, equal to a, or equal to $-a$. Then a_r is some integer times a^r, and Δ_s is some integer times a^p, where $p = \frac{1}{2}s(s+1)$. Stability does not depend on the magnitude of a but on whether all the integers are positive. This is related to the fact that for o.d.e. one can set down sign stability conditions, which were quoted in Section 6.3 and will be discussed in Section 8.1.

7.2 DIFFERENCE EQUATIONS

It must be strongly emphasized that for difference equations, in the case just discussed for o.d.e., the magnitude of a is significant. There are no sign stability conditions involving single coefficients of the secular equation for a system of coupled difference equations. This should not be surprising, since the primary stability condition is no longer $\mathrm{Re}(\lambda) < 0$, a sign condition, but $|\lambda| < 1$, a magnitude condition. However, it is worth thinking some more about the simplest equations, involving a single real parameter a

$$\frac{dx}{dt} = ax \tag{7.18}$$

and

$$x_{t+1} = ax_t \tag{7.19}$$

Equation (7.18) has solution $x_0 \exp(at)$, which tends to zero for $a < 0$, and to infinity for $a > 0$. On the other hand, equation (7.19) has solution $x_0 a^t$, which tends to zero for $|a| < 1$, and to infinity for $|a| > 1$. With reference to ordinary differential equations, one can talk of negative feedback as stabilizing and positive feedback as destabilizing. However, with reference to difference equations, one must talk of low gain as stabilizing and high gain as destabilizing. (See also Appendix 2.)

Before we examine in detail the stability procedure for a difference equation, another point is worth raising, which has been emphasized in particular by May (1973). It is often useful, in comparing models for species with overlapping generations and models for species with discrete generations, to set down certain o.d.e. and certain difference equations which can reasonably be regarded as analogous. This is not to be done by setting the a_{ij} identical in the two cases. For the derivative dx/dt is the analogue of a difference of two successive values of x_t. Thus the direct analogue of equation (7.18) is

$$x_{t+1} = ax_t + x_t \tag{7.20}$$

In general, on comparing two sets of first-order equations, each diagonal term a_{ii} in the o.d.e. model should be replaced by $a_{ii} + 1$, with the consequence that the analogous stability condition to $\mathrm{Re}(\lambda) < 0$, is $|\lambda + 1| < 1$. The two stability regions are compared in Fig. 7.4. It is obvious that the difference equation analogue always has a more stringent stability condition than does the original o.d.e. If there is reason to believe that only entry to the circle from the right

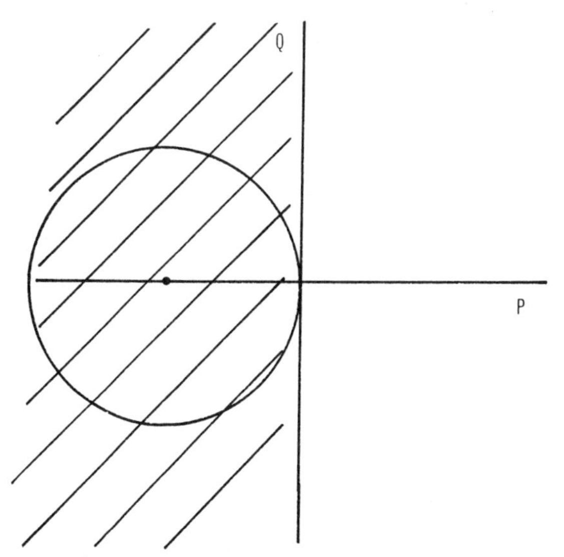

Fig. 7.4 The stability region (shaded) for an o.d.e., and that (the interior of the circle) for the analogous difference equation. The root is $P + iQ$, and the centre of the circle is at $P = -1$, $Q = 0$

matters, and that fluctuations with high-frequency components are not significant, then crossing the imaginary axis adequately approximates entering the circle. Otherwise one must use results that hold precisely for difference equations.

In what follows here, I am not concerned with this particular aspect of the comparison between o.d.e. and difference equation results, and I shall call diagonal elements a_{ii} as before. One can set down rather crude stability conditions from equations (7.10) alone. In the case $n = 4$ these are

$$\begin{aligned} -1 &< a_4 < +1 \\ -4 &< a_3 < +4 \\ -6 &< a_2 < +6 \\ -4 &< a_1 < +4 \end{aligned} \tag{7.21}$$

We cannot expect such independent conditions on the individual a_r to be particularly useful. For example, large negative values of a_1 come from having all λ_i positive, which will also give large negative values of a_3. It is necessary to find conditions on appropriate linear combinations of the a_r.

A convenient method for precise analysis of local stability is to find a change of variables $\lambda \to z$, such that $|\lambda| < 1$ corresponds to $\mathrm{Re}(z) < 0$. The original polynomial $P(\lambda)$ then gives a new polynomial $S(z)$ for which the 'positive coefficient' and 'Routh–Hurwitz' conditions can be used. A suitable transformation is

$$\lambda = \frac{z+1}{z-1}, \quad z = \frac{\lambda + 1}{\lambda - 1} \tag{7.22}$$

Let us now apply this to the case $n = 4$, by multiplying throughout by the factor $(z-1)^4$ to get the new polynomial

$$\begin{aligned} S(z) &= b_0 z^4 + b_1 z^3 + b_2 z^2 + b_3 z + b_4 \\ &= (z+1)^4 + a_1(z+1)^3(z-1) + a_2(z^2-1)^2 \\ &\quad + a_3(z+1)(z-1)^3 + a_4(z-1)^4 \end{aligned} \tag{7.23}$$

The conditions $b_r > 0$ give the following conditions on the original coefficients a_s

$$1 + a_1 + a_2 + a_3 + a_4 > 0 \tag{7.24a}$$
$$4 - 2a_1 + 2a_3 - 4a_4 > 0 \tag{7.24b}$$
$$6 - 2a_2 + 6a_4 > 0 \tag{7.24c}$$
$$4 + 2a_1 - 2a_3 - 4a_4 > 0 \tag{7.24d}$$
$$1 - a_1 + a_2 - a_3 + a_4 > 0 \tag{7.24e}$$

Adding equations (b) and (d) gives a condition on a_4, i.e. $a_4 < 1$, and adding equations (*a*), (*c*), and (*e*) gives another, $a_4 > -1$. Further manipulation gives linear inequalities for each of a_1, a_2, and a_3 separately in terms of a_4. The Routh–Hurwitz condition $\Delta_3 > 0$ has also to be applied. It should be noted that in the Routh–Hurwitz determinant Δ_s as defined by equation (7.12), 1 must be replaced by b_0 as well as a_r by b_r.

As a numerical problem for given values of the a_{ij} there is not much more to the local stability analysis of difference equations than to that of o.d.e. Many alternative methods can be used, in addition to that just described. When it is a matter of a qualitative analysis of the effect of changing some a_{ij}, or some parameter in the underlying non-linear equations, the results in equation (7.24) are a degree more opaque than the corresponding o.d.e. results, $a_r > 0$. Most of the applications of cycle analysis in the next chapter accordingly deal with o.d.e. models, although a difference equation model is used to illustrate rosette structure.

References

Gantmacher, F. R. (1959). *Applications of the theory of matrices.* Interscience.

Levins, R. (1974). The qualitative analysis of partially specified systems, *Proc. N.Y. Acad. Sci.* **231**, 123–138.

Levins, R. (1975a). Evolution in communities near equilibrium, in *Ecology and Evolution in Communities* (edited by M. L. Cody and J. M. Diamond), Harvard University Press, pp. 16–50.

Levins, R. (1975b). Problems of signed digraphs in ecological theory, in *Ecosystem Analysis and Prediction* (edited by S. A. Levin), Society for Industrial and Applied Mathematics, pp. 264–277.

May, R. M. (1973). *Stability and Complexity in Model Ecosystems*, Princeton Monographs in Population Biology 6, Princeton University Press.

Pitts, W. (1942). The linear theory of neuron networks: the static problem, *Bull. Math. Biophysics* **4**, 169–175.

Roberts, F. S., and Brown, T. A. (1975). Signed digraphs and the energy crisis, *Amer. Math. Monthly* **82**, 577–594.

Roberts, F. S. (1976). *Discrete Mathematical Models*, Prentice-Hall.

Tyson, J. J., and Othmer, H. G. (1978). The dynamics of feedback control circuits in biochemical pathways, *Prog. theor. Biol.* **5**, 1–62.

Chapter 8

Applications of cycle analysis

In this chapter I shall examine some applications of the local stability analysis outlined in the previous chapter, and in particular of cycle analysis in terms of equation (7.7). These applications are:

(1) sign stability analysis, following Quirk and Ruppert (1965), of a set of ordinary differential equations. A sufficient discussion will be given to indicate how it is possible to ensure local stability by specifying zero values or signs for the elements a_{ij} of the Jacobian matrix.
(2) a rough analysis, following Levins (1974), of the relative importance of predation and competition as destabilizing factors in a large community model formulated in terms of ordinary differential equations.
(3) a rough estimate, correcting that of Levins (1975), of the relative rates of return to equilibrium of stable, slowly oscillating systems. This makes use only of equations (7.10).
(4) cycle analysis of the destabilizing effect of distributed time delays in a model of a predator on two mutually competing prey species.
(5) cycle analysis of the Goodwin oscillator (Goodwin 1965, Tyson and Othmer 1978), which is a schematic model of an enzyme control loop.
(6) an illustration of rosette structure in a model of an annual plant population with a seedbank.

8.1 SIGN STABILITY

In sign stability analysis the aim is to find conditions on the elements a_{ij} of the Jacobian matrix which will ensure local stability, irrespective of the actual numerical values of the a_{ij}. Either certain a_{ij} are specified to be zero, or their signs are specified. The reason for seeking these results is that it is frequently very difficult in biological modelling, as in the economic modelling in which the approach originated, to pin down numerical values with any accuracy. But it may be relatively easy either to specify the signs of the interactions, or to indicate that certain interactions are negligible. The disadvantage of sign stability analysis is that the discovery that a system is sign unstable tells us rather little.

The procedure described by Quirk and Ruppert (1965) is to look successively at the increasingly complicated stability conditions of equations (7.9a), (7.9b), and (7.9c), and $\Delta_2 > 0, \ldots$ and so on. Any sign stability condition found at one step is fed into the next step. From equation (7.9a) the first condition is that all the a_{ii} are zero or negative, and that one at least is negative. Applying this condition in equation (7.9b), the second condition is that all products $a_{ij}a_{ji}$, $(i \neq j)$ are negative or zero. Equations (7.9c) and (7.13) focus attention on cycles of length 3. It is clearly going to be difficult to guarantee positive signs for both

$$a_3 = -\sum_{\substack{i\neq j, i\neq k\\ j\neq k}} a_{ij}a_{jk}a_{ki} + \sum_{\substack{i\neq j, i\neq k\\ j\neq k}} a_{ii}a_{jk}a_{kj} - \sum_{\substack{i\neq j, i\neq k\\ j\neq k}} a_{ii}a_{jj}a_{kk} \tag{8.1}$$

and

$$\Delta_2 = a_1 a_2 - a_3 = \sum_{\substack{i\neq j, i\neq k\\ j\neq k}} a_{ij}a_{jk}a_{ki} + \sum_{i} a_{ii} \sum_{j\neq i} a_{ij}a_{ji} - \sum_{i} a_{ii}^2 \sum_{j\neq i} a_{jj} \tag{8.2}$$

The two conditions already established make the second and third sums positive in both equation (8.1) and (8.2). In the second equation, in these sums, terms in the product $a_1 a_2$ cancel all terms in $-a_3$, leaving positive terms with repeated index i. But the first sum cannot be positive in both equations (8.1) and (8.2), so the third condition has to be that all the 3-cycles are zero. The full analysis by Quirk and Ruppert (1965) shows that all higher cycles must also be zero.

8.2 COMPETITION AND PREDATION

Equation (8.2) can be used to illustrate that in a large dimension model using ordinary differential equations, predation is less likely to lead to instability than is competition. However, when long cycles are not excluded, predation can have a destabilizing effect. To keep a close parallel with the model of May (1972) discussed in Section 6.2, let all the a_{ii} be equal to -1. Let a fraction C of the a_{ij} $(i = j)$ be non-zero. In the pure competition model all these are taken to be $-a$. In equation (8.2) the number of pairs i, j in the second and third sums is $\frac{1}{2}n(n-1)$. The number of 3-cycles is $\frac{1}{3}n(n-1)(n-2)$. So

$$\Delta_2 = \tfrac{1}{6}[2(n-2)(-Ca)^3 - 3(Ca)^2 + 3]\, n(n-1) \tag{8.3}$$

For any n, large enough Ca will destabilize, a result usually termed the Gause exclusion principle. For large enough n, any Ca will destabilize.

The pure predation model has, for any $a_{ij} = b$, a reciprocal $a_{ji} = -b$. The sum of 3-cycles will involve sign cancellations, but because of the random assignment of zero a_{ij} it will not necessarily be zero. So in this case

$$\Delta_2 = \tfrac{1}{6}[2(n-2)(\pm Cd)^3 + 3(Cb)^2 + 3]\, n(n-1) \tag{8.4}$$

where $d \ll b$. Instability can still occur for large enough n, but is much less likely than in the pure competition model.

8.3 A MEASURE OF RELATIVE STABILITY

In this section only stable cases are considered. All roots λ_i are either negative real, $\lambda_m = -p_m$, or complex with negative real part, $\lambda_k = -q_k \pm is_k$. The general solution of the linearized equations around the equilibrium point is a function of time t

$$X_j(t) = \sum_m A_{jm} \exp(-p_m t) + \sum_k B_{jk} \exp(-q_k t) \cos(s_k t)$$

$$+ \sum_k C_{jk} \exp(-q_k t) \sin(s_k t) \qquad (8.5))$$

The magnitude of any term here falls to $1/e$ of its original or highest value in a time p_m^{-1} or about q_k^{-1}. The mean rate of return to equilibrium may be defined as the reciprocal of the average value of these quantities

$$\rho = n/\left[\sum_m p_m^{-1} + 2\sum_k q_k^{-1}\right] \qquad (8.6)$$

Now we have no results for the q_k explicitly in terms of the coefficients a_r. If we did, stability theory would be a lot simpler. The best we can do is to use a result for the sum of the reciprocals of the roots, defining

$$\rho' = -n/\left[\sum_i \lambda_i^{-1}\right] \qquad (8.7)$$

We shall first see that this is an upper bound to ρ, and then that by using equations (7.10) it can be expressed in terms of the coefficients a_n and a_{n-1}. The sum of the reciprocals of the roots is, in this case when all real roots and real parts of roots are negative, given by

$$-\sum_i \lambda_i^{-1} = \sum p_m^{-1} - \sum_k \left[\frac{1}{(-q_k + is_k)} + \frac{1}{(-q_k - is_k)}\right]$$

$$= \sum_m p_m^{-1} + 2\sum_k \frac{q_k}{(q_k^2 + s_k^2)}$$

$$\leqslant \sum p_m^{-1} + 2\sum q_k^{-1} \qquad (8.8)$$

It follows that $\rho' \geqslant \rho$. For a slowly oscillating system, with all the s_k small compared with the q_k, ρ' is a reasonable approximate measure of the mean rate of return to equilibrium.

From equations (7.10), giving the coefficients a_r in terms of sums of products of the roots, it follows that the ratio of the final two coefficients is

$$\frac{a_{n-1}}{a_n} = -\Sigma\, \lambda_i^{-1} \tag{8.9}$$

So the approximate measure of the rate of return is

$$\rho' = \frac{n\, a_n}{a_{n-1}} \tag{8.10}$$

In the special case of a rosette this can be expressed as

$$\frac{\text{number of components} \times \text{summed weight of longest cycles}}{\text{summed weight of next longest cycles}}$$

8.4 DISTRIBUTED DELAYS

Cycle analysis should be particularly useful in comparing two systems between which the only essential difference is that one or more cycles present in the first are either absent, or of different length, in the second. Certain forms of delay in the interaction between two components of a system can be described in terms of additional components, so that the delayed effect of j on i is treated as if j acted on k, k on l, and so on until a final intermediate component s acted on i. So any cycle containing the arc (j, i) is transformed into a longer cycle containing the arcs $(j, k), (k, l), \ldots, (s, i)$. The qualitative analysis of the effect of this type of delay on stability should be facilitated by the use of equation (7.7).

A full treatment of this way of looking at delays is given in my earlier book (MacDonald 1978), which can be supplemented by the recent more rigorous treatment by Cooke and Grossman (1982). Let us start from a non-linear ordinary differential equation model in which one of the interaction terms $f_i(x_1, \ldots, x_j, \ldots, x_n)$ is replaced by a term including instead of the variable x_j a certain average over all earlier values of that variable

$$\bar{x}_j = \int_{-\infty}^{t} x_j(T)\, G_a^p(t - T)\, dT \tag{8.11}$$

with the memory function $G_a^p(u)$ given by a power factor and an exponential factor

$$G_a^p(u) = \frac{a^{p+1}\, u^p}{p!} \exp(-au) \tag{8.12}$$

From equations (8.11) and (8.12) the average delay $\bar{T}$ is given by

$$\bar{T} = (p + 1)/a \tag{8.13}$$

This change to a delayed variable can be carried out by

(*a*) using in place of the original f_i the form $f_i(x_1, \ldots, x_{n+p+1}, \ldots, x_n)$
(*b*) introducing a set of new linear equations

$$\frac{dx_{n+p+1}}{dt} = a(x_{n+p} - x_{n+p+1})$$

$$\frac{dx_{n+p}}{dt} = a(x_{n+p-1} - x_{n+p})$$

$$\cdots\cdots\cdots\cdots\cdots\cdots \qquad (8.14)$$

$$\frac{dx_{n+1}}{dt} = a(x_j - x_{n+1})$$

In the linear stability analysis around the equilibrium point

$$(x_1^0, x_2^0, \ldots, x_j^0, \ldots, x_n^0, x_j^0, \ldots, x_j^0)$$

the Jacobian matrix is altered, from its form in the model without delay, in the following way:

(*a*) a_{ij} becomes zero
(*b*) all other a_{kl} for k, l from 1 to n are unchanged.
(*c*) $p + 1$ new rows and columns are added, with zero elements except

$$a_{rr} = a, \; r = n + 1 \text{ to } n + p + 1$$

$$a_{rr-1} = a, \; r = n + 2 \text{ to } n + p + 1$$

$$a_{n+1j} = a$$

$$a_{in+p+1} \text{ taking the original value of } a_{ij}$$

As a result, any cycle containing $a_{ij} = b$ must be replaced by a longer cycle containing the product

$$a_{i\,n+p+1}\, a_{n+p+1\,n+p}\, a_{n+p\,n+p-1} \cdots a_{n+2\,n+1}\, a_{n+1\,j} = b\, a^{p+1} \qquad (8.15)$$

Also there are $p + 1$ new 1-cycles of weight $-a$. Here there are two conflicting effects. One is the removal of a cycle of length r and its replacement by one of length $r + p + 1$. This is likely to be destabilizing either because of a change in sign of a coefficient a_r, or more typically because of the way long and short cycles enter the Routh–Hurwitz determinants. The other effect is a stabilizing one, the addition of new 1-cycles of negative weight.

Fig. 8.1 shows the digraph of a model with three components when delays of this kind, with $p = 1$, are inserted in two of the interactions. In place of two cycles of length 2 there are now two cycles of length 4. Previously there were two stabilizing 1-cycles, now there are six. A particular example of this is a linearized model of a predator species with two prey species, which are competitors and which have self-stabilizing intraspecies competition. The influence of either prey population on the rate of increase of the predator population is likely to be

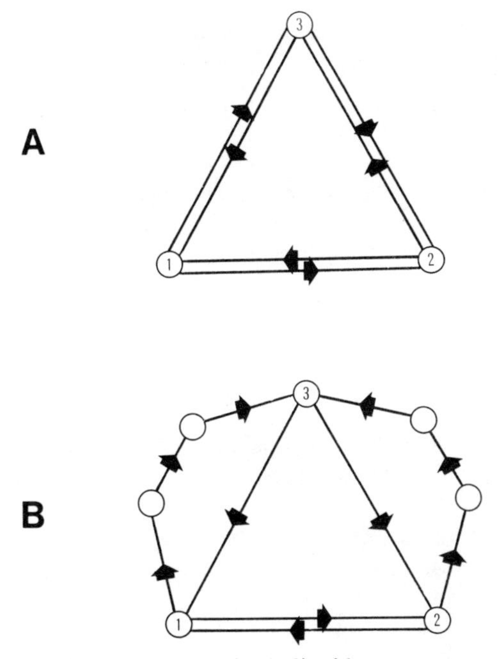

Fig. 8.1 A predator (node 3) with two competing prey. In **A** there is no delay, while in **B** each of the links from prey to predator is subject to a distributed delay described by two additional linear o.d.e.

delayed. The two Jacobians are

$$\begin{vmatrix} a_{11} & a_{12} & a_{13} \\ a_{21} & a_{22} & a_{23} \\ a_{31} & a_{32} & a_{33} \end{vmatrix} = \begin{vmatrix} -b_{11} & -b_{12} & -b_{13} \\ -b_{21} & -b_{22} & -b_{23} \\ b_{31} & b_{32} & 0 \end{vmatrix} \tag{8.16}$$

where the b_{ij} are positive, and

$$\begin{vmatrix} -b_{11} & -b_{12} & -b_{13} & 0 & 0 & 0 & 0 \\ -b_{21} & -b_{22} & -b_{23} & 0 & 0 & 0 & 0 \\ 0 & 0 & 0 & b_{31} & 0 & b_{32} & 0 \\ 0 & 0 & 0 & -a & a & 0 & 0 \\ a & 0 & 0 & 0 & -a & 0 & 0 \\ 0 & 0 & 0 & 0 & 0 & -a & a \\ 0 & a & 0 & 0 & 0 & 0 & -a \end{vmatrix} \tag{8.17}$$

in which the 2-cycles with weights $-b_{13}b_{31}$ and $-b_{23}b_{32}$ are replaced by 4-cycles with weights $-a^2 b_{13} b_{32}$ and $-a^2 b_{23} b_{32}$.

In the absence of delay the coefficient a_2 in the secular equation is given, according to equation (7.7), by

$$(b_{13}b_{31} + b_{23}b_{32} - b_{12}b_{21} + b_{11}b_{22}) \tag{8.18}$$

Now let us assume that the system is stable, so that the stabilizing effect of the two predation cycles, together with the second-order effect of the two loops, outweighs the destabilizing effect of interspecies competition. Let us also assume that the second-order effect on its own is not enough for stability

$$b_{12}b_{21} > b_{11}b_{22}$$

When the delays are included the coefficient a_2 becomes

$$(-b_{12}b_{21} + b_{11}b_{22} + 2a(b_{11} + b_{22}) + 6a^2) \tag{8.19}$$

If the mean delay is long enough, that is if a is small enough, a_2 will be negative, and the delay destabilizes the equilibrium.

If on the other hand $b_{12}b_{21} < b_{11}b_{22}$, the coefficient a_2 remains positive, and it is necessary to go through the Routh–Hurwitz procedure to test for instability. This is a more typical outcome of including a delay, giving an oscillatory instability.

8.5 THE GOODWIN OSCILLATOR

The enzymes in biochemical pathways leading to certain essential metabolites, such as nucleic acids, are synthesized in the cell in the absence of an external supply of the metabolite. When the metabolite is added to a growth medium in which the cells are immersed, synthesis of these enzymes is repressed. It is natural to consider that these pathways are controlled by a negative feedback dependent on the concentration of the end-product metabolite. It is known that several steps can be present in the pathway, for example, in the synthesis of arginine, histidine and lysine as many as eight or nine. A schematic model of a pathway with end-product inhibition, due in the first instance to Goodwin (1965), makes all stages proceed according to linear rate equations, except the production of the first intermediate enzyme. This is taken to depend on the concentration of the end-product in a specific non-linear manner, as a Hill function

$$f(x_n) = [1 + a\,x_n^{\rho}]^{-1} \tag{8.20}$$

The underlying model of inhibition leading to equation (8.20), and the interpretation of the parameter ρ, requires ρ molecules of the end product S_n to react with an enzyme E to give an inactive form S

$$E + \rho S_n \rightleftharpoons S$$

E catalyses production of the first intermediate product S_1, but S cannot do this. A value of 2 or more for ρ implies a cooperative aspect to the inhibition.

To conform to notations used in Chapter 9, and to allow an explicit solution for the roots of the secular equation, I shall take the model in the form

$$\frac{dx_1}{dt} = \frac{\theta^\rho}{\theta^\rho + x_n^\rho} - k x_1$$

$$\frac{dx_r}{dt} = k(x_{r-1} - x_r), r = 2, \ldots, n \tag{8.21}$$

There is a strong resemblance to a distributed delay model here, and indeed the sequence of reactions, if long enough, has a destabilizing effect akin to that of a delay. Linearization about the equilibrium point $(x, x, \ldots, x)$ gives a Jacobian matrix with all $a_{ij} = 0$ except for

$$a_{ii} = -k, \ i = 1, \ldots, n$$

$$a_{rr-1} = k, \ r = 2, \ldots, n \tag{8.22}$$

$$a_{1n} = -y = -\frac{\rho\theta^\rho x^{\rho-1}}{(\theta^\rho + x^\rho)^2}$$

Here the equilibrium value x of each x_i is the solution of

$$kx = \theta^\rho/(\theta^\rho + x^\rho) \tag{8.23}$$

The digraph structure is displayed in Fig. 7.1. There are n 1-cycles of weight $-k$, and a single n-cycle with weight $-k^{n-1}y$.

The secular equation takes the form

$$(\lambda + k)^n + k^{n-1}y = 0$$

All the coefficients a_r are positive, and stability can be tested using the Routh–Hurwitz determinants. However, here we have the rare opportunity of by-passing this cumbersome process by finding the roots explicitly. It is convenient to tidy up the secular equation by writing $y = kz$, giving

$$z = \frac{\rho x^\rho}{\theta^\rho + x^\rho} \tag{8.24}$$

and the secular equation in the form

$$(\lambda + k)^n + k^n z = 0 \tag{8.25}$$

Equation (8.25) has roots

$$\lambda_m = -1 + kz^{1/n} \exp\left(\frac{i\pi}{n} + \frac{2im\pi}{n}\right)$$

and of these the two which are associated with instability are

$$\lambda_0, \lambda_{n-1} = -k + kz^{1/n} \exp\left(\pm\frac{\pi i}{n}\right) \tag{8.26}$$

Table 8.1 Necessary values of the steepness parameter ρ for instability of the equilibrium point of the Goodwin oscillator chain having n steps. The parameter ρ takes the value $\cos^{-n}(\pi/n)$ at the onset of instability

n	ρ
2	∞
3	8
4	4
6	64/27
∞	1

The critical value of z, at which these two roots take their pure imaginary values $\pm i\omega_0 = \pm ik\tan(\pi/n)$, is

$$z = \cos^{-n}(\pi/n) \tag{8.27}$$

Now ρ by the definition of equation (8.24) exceeds z, so that instability is only possible for values of ρ that are at least as great as $\cos^{-n}(\pi/n)$. The implications are displayed in Table 8.1. Unless there are very long chains, oscillatory instability only comes about for quite a high degree of cooperativity of the end-product molecules in the inhibition.

8.6 A SEEDBANK PROBLEM, AND SOME OBSERVATIONS ON ROSETTES

Roberts and Brown (1975) found that rosette structures appeared in some of their models for environmental and urban problems. They are of limited applicability in biological models, such as those of interacting species populations. There are two main reasons for this, the prevalence of loops and the reciprocity of most interactions. In defining a rosette in Section 7.1 I noted that at most one loop is allowed, at the common node. But in population problems self-stability is certainly a feature of the populations at the lowest trophic level, and as mentioned in Section 6.3 it may be more general.

If all the components have a self-stable loop with the a_{ii} all equal, then formally one can use the rosette structure with λ replaced by $\lambda + a$, where a is the common value of the a_{ii}. This device also allows the qualitative assessment of time delay effects, using the distributed delay of equation (8.12), holding a fixed and obtaining the required $\bar{T}$ by suitable choice of p. However, one finally has to come down to using the Routh–Hurwitz analysis, and the interpretation in terms of the original a_{ij} is roundabout. I have discussed this aspect of rosette analysis elsewhere (MacDonald 1977).

In the models discussed by Roberts and Brown (1975), one-way effects are common. In population interactions these may be more prevalent than has previously been thought (Lawton and Hassell 1981), but it is typically the case that an interaction term a_{ij} is associated with an a_{ji} of comparable magnitude. So in this context a model is more likely to have a digraph such as **A** in Fig. 8.2 than a rosette digraph such as **B**.

For a specific rosette example I turn to the problem of an annual plant population with a seedbank. It is frequently found that the seeds deposited one year by an annual plant do not all germinate the following year. A sizeable fraction of them can remain in the soil for several years, still being capable of germination. The population dynamics of an annual species has been modelled (MacDonald and Watkinson 1981) in terms of two populations satisfying coupled difference equations. These are the population of mature plants, and the population of seeds in the bank. In its simplest form, ignoring the fact that seeds can die or be eaten while in the bank, such a model takes the form

$$\begin{aligned} N_{t+1} &= a\,F(N_t) + b\,S_t \\ S_{t+1} &= (1-a)\,F(N_t) + (1-b)\,S_t \end{aligned} \tag{8.28}$$

It is assumed that the only non-linear effect, expressed through the form chosen for $F(N_t)$, is in the production of seeds. The non-linear self-thinning effect in dense populations of seedlings is ignored. For our present purposes, the other significant approximation in equations (8.28) is that any seed, once it has joined the seedbank, is equally likely to germinate irrespective of its age.

There is some evidence that the probability of germination falls off with the age of the seed. This suggests the use of a more elaborate model, of the form

$$N_{t+1} = a\,F(N_t) + (1-a)[b_1F(N_{t-1}) + b_2F(N_{t-2}) + \cdots + b_nF(N_{t-n})] \tag{8.29}$$

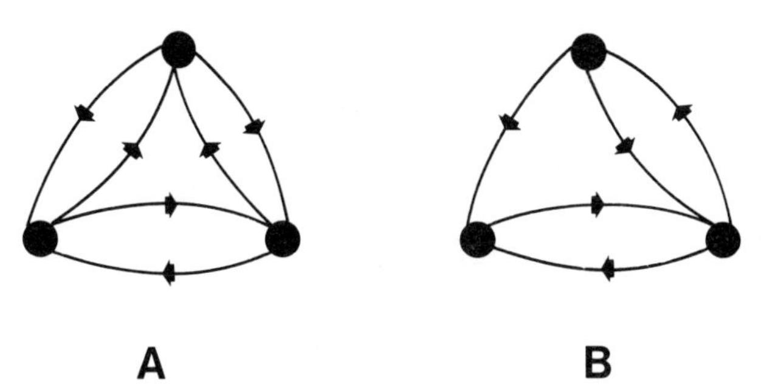

Fig. 8.2 **A** shows three species with all interactions reciprocal, such as competition, mutualism or predation with Lotka–Volterra dynamics. In **B** one interaction is made one-way, for example by replacing the mutualistic relationship by amensalism, or taking predation to be donor controlled. In **B** the digraph is a rosette

Again, one may assume that no seeds are lost from the bank except by germination, so that the sum of the b_i is 1. Then the equilibrium population N^0 is the same as it would be in the absence of the bank

$$N^0 = F(N^0)$$

With the positive quantity c defined by

$$\left.\frac{\partial F(N)}{\partial N}\right|_{N=N^0} = -c$$

the linearized equation for $X_t = N_t - N^0$ becomes

$$X_{t+1} = -c[aX_t + (1-a)(b_1X_{t-1} + b_2X_{t-2} + \cdots + b_nX_{t-n})] \quad (8.30)$$

This difference equation of order $n+1$ can be written as a set of $n+1$ first-order equations

$$\begin{aligned} X^1_{t+1} &= -c[aX_t + (1-a)\,b_1X^2_t] \\ X^2_{t+1} &= X^1_t + (b_2/b_1)X^3_t \\ X^3_{t+1} &= X^1_t + (b_3/b_2)\,X^4_t \\ &\cdots\cdots\cdots\cdots\cdots\cdots \\ X^n_{t+1} &= X^1_t \end{aligned} \quad (8.31)$$

The digraph for these linear equations is a rosette, as displayed in Fig. 8.3. In the secular equation one term comes from each cycle in the rosette, so that

$$\lambda^{n+1} + ca\lambda^n + (1-a)\,c[b_1\lambda^{n-1} + b_2\lambda^{n-2} + \cdots + b_{n-1}\lambda + b_n] = 0 \quad (8.32)$$

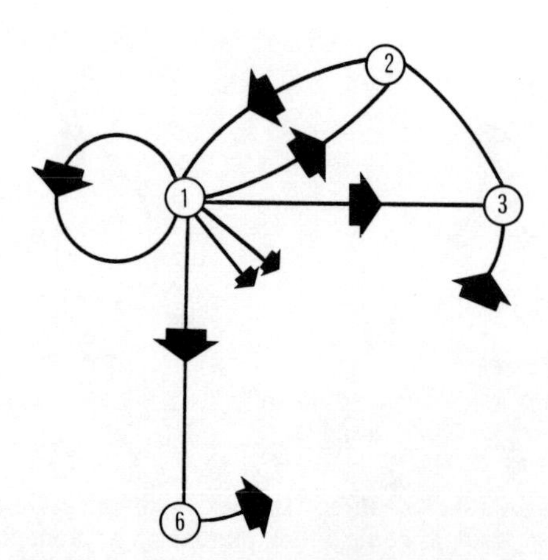

Fig. 8.3 The rosette digraph for the seedbank model in the case $n = 5$. There are cycles of length 1, 2, 3, ... with one common node

References

Cooke, K. L., and Grossman, Z. (1982). Discrete delay, distributed delay and stability switches, *J. math. Anal. App.* **86**, 592–627.

Goodwin, B. C. (1965). Oscillatory behaviour in enzyme control processes, *Adv. Enzyme Reg.* **3**, 425–438.

Lawton, J. H., and Hassell, M. P. (1981). Asymmetrical competition in insects, *Nature* **289**, 793–795.

Levins, R. (1974). The qualitative analysis of partially specified systems, *Ann. N.Y. Acad. Sci.* **231**, 123–138.

Levins, R. (1975). Problems of signed digraphs in ecological theory, in *Ecosystem Analysis and Prediction* (edited by S. A. Levin), Society for Industrial and Applied Mathematics, Philadelphia, pp. 264–277.

MacDonald, N. (1977). Time lag in systems described by difference or differential equations, *Int. J. Systems Sci.* **8**, 467–479.

MacDonald, N. (1978). *Time Lag in Biological Models*, Lecture Notes in *Biomathematics* 27, Springer, Berlin.

MacDonald, N., and Watkinson, A. R. (1981). Models of an annual plant population with a seedbank, *J. theor. Biol.* **93**, 643–653.

May, R. M. (1972). Will a large complex system be stable?, *Nature* **238**, 413–414.

Quirk, J. P., and Ruppert, R. (1965). Qualitative economics and the stability of equilibrium, *Rev. Econ. Stud.* **32**, 311–326.

Roberts, F. S., and Brown, T. A. (1975). Signed digraphs and the energy crisis, *Amer. Math. Monthly* **82**, 577–594.

Tyson, J. J., and Othmer, H. F. (1978). The dynamics of feedback control circuits in biochemical pathways, *Prog. theor. Biol.* **5**, 1–62.

Chapter 9

Power laws and switching functions

In the previous two chapters I have looked, within a linear approximation, at the consequences of the cycle structure of the interactions between the components of a complex system. Once linearity is abandoned there are still possibilities for gaining insight into the general consequences of cycle structure. In this chapter I shall examine two particular forms of non-linearity in models formulated in terms of ordinary differential equations.

These methods can be illustrated in terms of a non-linear function already encountered. This is the inhibition function used in the Goodwin model (as described in Section 8.5)

$$F(x) = \frac{\theta^{\rho}}{\theta^{\rho} + x^{\rho}} \tag{9.1}$$

I shall consider two approximations to this function:

(*a*) power law

$$F(x) = (\theta/x)^{\rho}$$

(*b*) switching function

$$\begin{aligned} F(x) &= 1,\ x < \theta, \\ &= 0,\ x > \theta \end{aligned}$$

The first may be adequate for large x but is misleading for small x. The second improves as ρ increases. The parameter θ becomes the threshold for the variable x. The emphasis in using a power law approximation is in making it easier to assess the sensitivity of the equilibrium values, and of the local stability properties, to variation of parameters in the rate equations. With switching functions the emphasis is on global behaviour.

9.1 POWER LAWS

A linear set of rate equations has the immediate advantage that the equilibrium values are easily found. So long as the matrix A is non-singular, the equations

$$\frac{d\boldsymbol{X}}{dt} = A\,\boldsymbol{X} + \boldsymbol{B} \tag{9.2}$$

have the unique equilibrium point

$$\boldsymbol{X}^0 = -A^{-1}\boldsymbol{B} \tag{9.3}$$

The Lotka–Volterra equations, discussed in Section 6.1, owe much of their popularity to the fact that their various equilibrium points can once more be found by matrix inversions. These equations

$$\frac{dx_i}{dt} = x_i\left(r_i + \sum_j a_{ij}x_j\right), i = 1, \ldots, n \tag{9.4}$$

have several equilibrium points:

(1) all x_i zero
(2) no x_i zero, equilibrium point given by

$$\boldsymbol{X}^0 = -A^{-1}\boldsymbol{r} \tag{9.5}$$

(3) one or more of the x_i zero, the others given by equations of the form of equation (9.5) but with the vector **r** and matrix A suitably truncated.

In situations where the Lotka–Volterra equations have been tried and seem to fail, a variety of minor modifications have been proposed. For example, Ayala *et al.* (1973) enumerate ten sets of equations to describe two-species competition. One of these is

$$\frac{dx_i}{dt} = x_i\left[r_i + \sum_j b_{ij}\ln(x_j)\right] \tag{9.6}$$

which can be exactly linearized by the transformation $y_i = \ln(x_i)$ but has little else to recommend it. Seven of the other sets involve powers of the x_i.

Savageau (1976) discusses the behaviour of equilibria in biochemical pathways using power law rate equations in a systematic manner. All his rate equations involve a production term and a decay term

$$\frac{dx_i}{dt} = a_i \prod_j (x_j)^{g_{ij}} - b_i \prod_j (x_j)^{h_{ij}} \tag{9.7}$$

The equilibrium point with all the x_i not zero can be found by a logarithmic transformation and a matrix inversion. For the equilibrium conditions are

$$(a_i/b_i) = \prod_j (x_j^0)^{h_{ij}-g_{ij}} \tag{9.8}$$

which are equivalent to the linear form

$$\ln(a_i) - \ln(b_i) = \sum_j \ln(x_j^0)\,(h_{ij} - g_{ij}) \tag{9.9}$$

As an example of the application of this method, Savageau sets up equations of the form of equation (9.8) for a model with the same structure as the Goodwin model

$$\frac{dx_1}{dt} = a_1 x_n^{g_{1n}} - b_1 x_1^{h_{11}} \tag{9.10}$$

$$\frac{dx_r}{dt} = a_r x_{r-1}^{g_{r,r-1}} - b_r x_r^{h_{rr}}, r = 2, \ldots, n \tag{9.11}$$

All the powers except g_{1n} are positive, while g_{1n} is negative to express the inhibition of the production of x_1 by the presence of the end product x_n. By making approximations similar in spirit to those described in Section 8.5, Savageau arrives at a condition that $-g_{1n}$ must exceed a quantity proportional to $\cos^{-n}(\pi/n)$ for local instability. So again there is a trade-off between length n of the chain of reactions and a parameter expressing the sharpness of inhibition. He integrates equations (9.11) numerically and obtains oscillations of growing amplitude in the unstable regime. This is no advance on the use of a linear approximation, the original Goodwin equations having a limit cycle solution in the unstable regime (Hastings *et al.* 1977).

Savageau (1976) discusses the sensitivity and stability of equilibrium for a variety of models of biochemical pathways with inhibition or activation by an end-product or an intermediate product, or with cascade activation. In later work he discusses allometric growth laws under the assumption that the underlying metabolic processes are governed by equations of the form of equation (9.8) (Savageau 1979).

The full advantages of power law rate equations appear when the powers are not merely fitted to whatever data may be available but are constrained stoichiometrically, as in chemical reaction kinetics. Here the rate equations contain powers of the concentrations of various molecular species. Because of the conservation of the atomic species, the powers appearing in the different rate equations for a given sequence of reactions must take consistent values. Probably the most general and powerful way of exploiting stoichiometry is that developed by Clarke. He has set this out in a long but tersely written review (Clarke 1980).

9.2 SWITCHING FUNCTIONS

I now turn to the switching function approach, and in particular to the work of Glass on systems of from two to five components with rate equations

$$\frac{dx_i}{dt} = \lambda_i B_i(\tilde{x}_j, j \neq i) - \gamma_i x_i \tag{9.12}$$

In these equations the $\tilde{x}_j$ are Boolean variables and the functions B_i are Boolean functions. A Boolean variable takes values 0 or 1. A Boolean function of a set of Boolean variables takes values 0 or 1 in a manner prescribed by a truth table or by writing the function in terms of certain basic Boolean functions. In the present example the Boolean variable $\tilde{x}_j$ is defined by

$$\begin{aligned} \tilde{x}_j &= 0, \ x_j < \lambda_j/\gamma_j = \theta_j \\ &= 1, \ x_j > \theta_j \end{aligned} \tag{9.13}$$

As a particular illustration of equations (9.12) take the analogue of the Goodwin model, in which production of x_1 is switched off when x_n exceeds its threshold value θ_n, and in which production of each of the other x_r is switched on when x_{r-1} exceeds θ_{r-1}. Writing the equations in the form of equation (9.12) requires that

$$\begin{aligned} B_1(\tilde{x}_n) &= 0 \quad \text{when} \quad \tilde{x}_n = 1 \\ &= 1 \quad \text{when} \quad \tilde{x}_n = 0 \end{aligned}$$

and

$$\begin{aligned} B_r(\tilde{x}_{r-1}) &= 1 \quad \text{when} \quad \tilde{x}_{r-1} = 1 \\ &= 0 \quad \text{when} \quad \tilde{x}_{r-1} = 0 \end{aligned}$$

In truth table form these functions are

$\tilde{x}$	$B_1(\tilde{x})$	$B_r(\tilde{x})$
0	1	0
1	0	1

They can also be given explicitly in terms of $\tilde{x}$ and the Boolean function

$$\bar{\tilde{x}} = \text{NOT } \tilde{x} = 1 - \tilde{x}$$

as

$$B_1(\tilde{x}) = \bar{\tilde{x}}, \ B_r(\tilde{x}) = \tilde{x}$$

In the present context the Boolean function B_1 is the switching analogue of the non-linear function $F(x)$ given by equation (9.1), while B_r is the switching analogue of

$$1 - F(x) = \frac{x^\rho}{\theta^\rho + x^\rho}$$

The thresholds θ_j divide the positive region of the n-dimensional space of the continuous variables into 2^n regions, each of which can be labelled by the appropriate values of the Boolean variables $\tilde{x}_j$, that is by a vector of 0s and 1s. So the 2^n regions are labelled as if they were the 2^n vertices of a hypercube. In this 'coded' description of the space, any trajectory of the solutions of equation (9.12) that passes from one region to another, as one variable crosses its threshold, is coded as an arc from one vertex to one of the adjacent ones. Here adjacent means that one and only one of the components of the labelling vector changes from 0 to 1 or from 1 to 0.

In this coded description a cyclic path is a sequence of arcs, each from a vertex to an adjacent vertex, with the last arc returning to the first vertex, but no other vertex being visited twice. A stable cyclic path is a cyclic path for which all the vertices on the path are the end-points of arcs from adjacent vertices not on the path. An n-dimensional cyclic path is a cyclic path on the n-cube that is not confined to a sub-cube. Three problems are examined in the work of Glass (Glass 1977a, Glass and Pasternack 1978a,b). The first is a purely combinatorial or graph-theoretical one, the enumeration of the n-dimensional stable cyclic paths for n up to five. Once these paths are identified the corresponding set of rate equations (9.12) can be set up, and the second problem is the numerical and analytical investigation of the nature of the trajectories of solutions of these equations. Finally, for each set of rate equations (9.12) with discontinuous non-linear functions, there corresponds a set with the smoothed out versions such as $F(x)$ of equation (9.1). It is necessary to examine whether the trajectories have the same qualitative behaviour in the discontinuous and smoothed out versions, since presumably no attainable biochemical switching process is truly discontinuous.

It is worth while at this point looking once more at the way the hypercube is used in Section 2.3, and isolating the similarities and distinctions between that application and the present one. In Landau's Markov process model, the decisions which adjacent vertex is available and whether to go to it are made according to probabilistic rules. Thus all pairwise encounters are equally likely, and specific hypotheses are made on the probability of a reversal of dominance in each possible encounter. The outcome is a distribution of the relative frequency of occupation of the various vertices. The models investigated by Glass are entirely deterministic. Since there are a finite number of vertices, either a steady state is reached (all paths lead to a single vertex, or divide into subsets of paths each leading to a single vertex) or there is a stable cyclic path.

In each of these hypercube applications, and for similar reasons, the only changes of vertex allowed are between adjacent vertices. In the Landau model encounters are assumed to be ordered in time, so changes in the relative status of pairs are ordered in time, and any such change means going to an adjacent vertex. In the Glass equations (9.12), although derivatives change discontinuously at a threshold, the variables x_j change continuously. We can ignore the possibility that two variables cross their thresholds simultaneously, and take threshold crossings as ordered in time. Again this means that an arc can only be from a vertex to an adjacent one. The intervals between two crossings may sometimes be extremely short but this has no bearing on the combinatorial aspect of the problem.

Finally there is a technical distinction between the dynamic digraphs used in Chapter 2 and those used in the present context. In Landau's model the state of the whole system was defined in terms of the states of the pairwise interactions—hence the term dynamic interaction digraph. Here the state of the whole system is defined in terms of the states (0 or 1) of the individual components. This may be called a **dynamic component digraph**.

9.3 STABLE CYCLES

Some hints to the answers to the three problems mentioned above are provided by the Goodwin model already studied. Firstly it suggests that in any dimension from $n = 3$ up there is at any rate one stable cyclic path. Secondly it indicates that the trajectory corresponding to this path in the version of the model with a smoothed out switching function will be either a limit cycle—for high enough dimension and sharp enough change of the switching function around threshold—or failing these conditions a decaying oscillation to an equilibrium point. Notice that in the continuous model a stable limit cycle is associated with an unstable equilibrium point, but there is no need for this concept in the combinatorial model.

Before proceeding to details of the enumeration of stable cyclic paths I shall summarize the main results obtained. Glass (1977a,b) finds two new stable 4-dimensional paths and seventeen new 5-dimensional ones, in addition to those of the Goodwin type. Glass and Pasternack (1978a) prove that when there is a stable cyclic path the trajectories of solutions of equations (9.12) are either stable limit cycles or decaying oscillations. Glass (1977a,b) verifies numerically that limit cycles occur in 4- and 5-dimensional cases not of the Goodwin type. Glass and Pasternack (1978b) verify numerically that these also occur in the smoothed out versions of these models. So there has been a considerable extension of the repertoire of limit cycles that may play a role in biochemical pathways.

Let us now look more closely at models of the type defined by equations (9.12) in two, three and four dimensions. In two dimensions the equations are

$$\frac{dx_1}{dt} = \lambda_1 B_1(\tilde{x}_2) - \gamma_1 x_1$$

$$\frac{dx_2}{dt} = \lambda_2 B_2(\tilde{x}_1) - \gamma_2 x_2 \tag{9.14}$$

The hypercube is simply a square with its vertices labelled 00, 01, 11, and 10. The unique stable cyclic path follows the vertex sequence $00 \rightarrow 10 \rightarrow 11 \rightarrow 01$ with the threshold crossings 1212. This case requires

$$B_1 = \bar{\tilde{x}}_2, \; B_2 = \tilde{x}_1$$

The smoothed out (Goodwin) version of this has decaying oscillations only.

In three dimensions the unique stable cyclic path is as shown in Fig. 9.1. It is labelled 000,100,110,111,011,001,000, or 123123. This is the Boolean analogue of the Goodwin model. Arcs from 101 and 010, the only two vertices not on the cyclic path, go to vertices on the cyclic path. In the original form of the Goodwin model (Tyson 1975), when the condition for local instability is satisfied, trajectories from the regions corresponding to these vertices either go to the equilibrium point or enter one of the six other regions and then approach the limit cycle.

It is possible to display the stable cyclic paths in four dimensions in a 2-dimensional diagram which looks like a cube inside a cube (Glass 1977a).

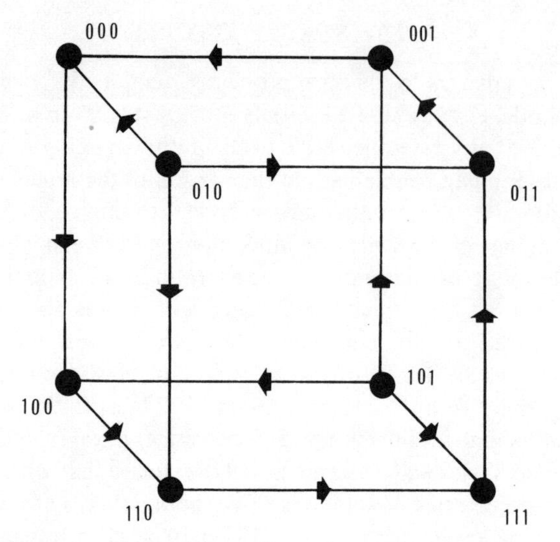

Fig. 9.1 The stable cyclic path on the cube (the Boolean analogue of the three-dimensional Goodwin oscillator) avoids the vertices 010 and 101. Arrows on the edges contiguous to these vertices point away from 010 or 101 and towards a point on the stable cyclic path. This path has length six

Another method will be used here. Fig. 9.2 shows the adjacent pairs of vertices connected by lines. Fig. 9.3 shows the Goodwin analogue stable cyclic path 12341234, and the paths 12314234, and 12314324. These are all the stable cyclic paths. The combinatorial methods used to identify these paths and the eighteen 5-dimensional stable cyclic paths are described in Glass (1977b).

Glass and Pasternack (1978b) give the explicit Boolean rate equations (9.12) and their continuous analogues for the paths 12314234 and 12314324. Unlike the Goodwin model, these involve inhibition by more than one component. The first is a symmetric set of equations

$$\frac{dx_i}{dt} = \bar{\bar{x}}_{i+1}\bar{\bar{x}}_{i+2} - x_i \tag{9.15}$$

or

$$\frac{dx_i}{dt} = \frac{\theta_{i+1}^{\rho}}{\theta_{i+1}^{\rho} + x_{i+1}^{\rho}} \frac{\theta_{i+2}^{\rho}}{\theta_{i+2}^{\rho} + x_{i+2}^{\rho}} - x_i \tag{9.16}$$

where as indices $5 = 1$ and $6 = 2$. A linearized version of this set of interactions can be displayed in the kind of digraph used in Chapters 7 and 8. This is given in Fig. 9.4. In this case, as for the Goodwin oscillator, there is a critical ρ for local

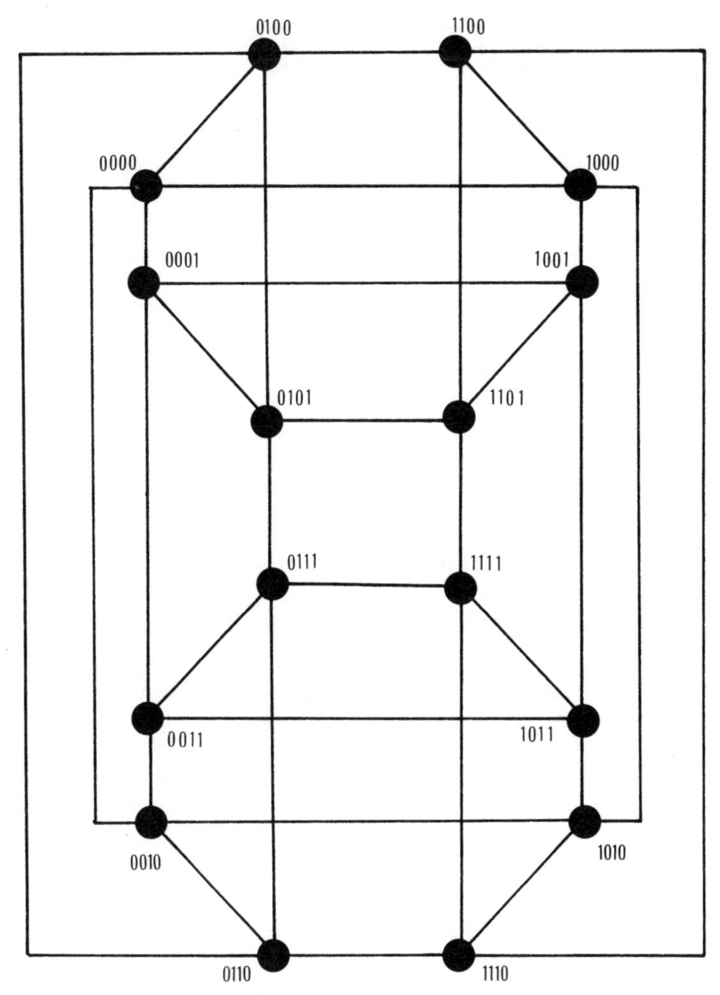

Fig. 9.2 The vertices of the hypercube in four dimensions (the 4-cube) with adjacent vertices linked. Reproduced from *Problems and Solutions in Logic Design*, by D. Zissos, by permission of the Oxford University Press. © 1976 by the Oxford University Press

instability of the equilibrium point ($x_i = \theta_i$) and a limit cycle is found when ρ exceeds this value.

In view of the symmetry in equations (9.15) and (9.16), it is not surprising that this pattern of interactions has been used previously in an oscillation model. Kling and Szekely (1968) proposed a double inhibition Boolean model with four components to describe the firing sequence of four neurons involved in swimming activity in the leech *Hirudo medicinalis*. The model agrees only in part with the

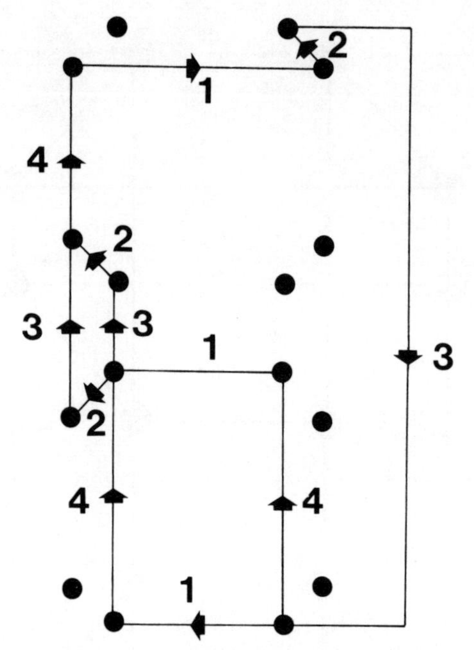

Fig. 9.3 The stable cyclic paths on the 4-cube. The vertices are to be taken as labelled in the same way as in Fig. 9.2. The paths are 12341234 (again the analogue of the Goodwin oscillator), 12314234 and 12314324

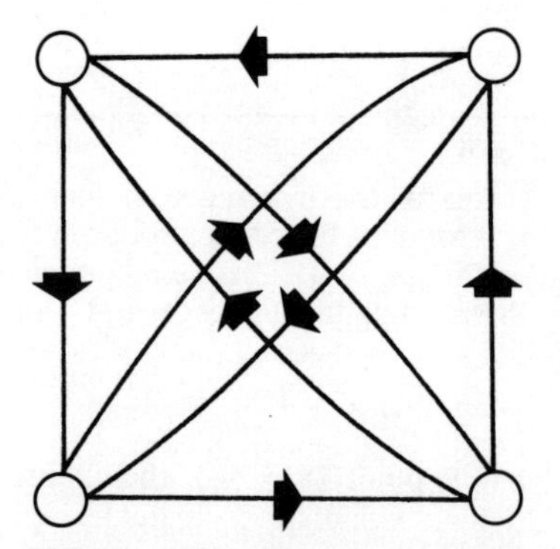

Fig. 9.4 The linearized representation of the interactions in an o.d.e. model corresponding to the stable cyclic path 12314234. All the interactions have negative sign, and there are negative loops at each node (not drawn)

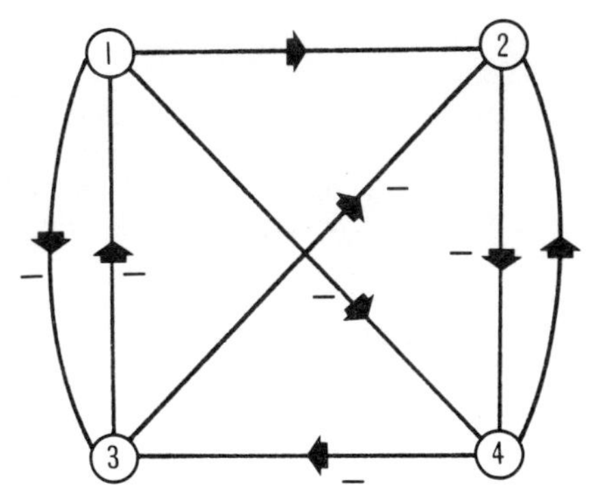

Fig. 9.5 The linearized representation of the interactions in an o.d.e. model giving the stable cyclic path 12314324. Negative-sign arcs are shown but negative loops are not drawn

data of Friesen *et al.* (1976) on the inhibitory connections and firing sequence of these neurons.

The final stable cyclic path relates to a less symmetric model—its linearized form is shown in Fig. 9.5—and it could hardly have been 'intuitively' hit upon as a likely model for an oscillator. It seems fair to say that no one is likely to pursue this combinatorial approach beyond $n = 5$. In the next chapter I shall examine models which abandon the use of ordinary differential equations and the ordering of level crossings in order to go to higher dimensions. In these models, which may be called clocked switching networks, components are Boolean switches operating synchronously. Small Boolean networks which are not clocked are studied from a quite different viewpoint from that of Glass in the volume edited by Thomas (1979).

References

Ayala, F. J., Gilpin, M. E., and Ehrenfeld, J. G. (1973). Competition between species; theoretical models and experimental tests, *Theor. Popn. Biol.* **4**, 331–356.

Clarke, B. L. (1980). Stability of complex reaction networks, *Adv. Chem. Phys.* **43**, 1–216.

Friesen, W. O., Poon, M., and Stent, G. S. (1976). An oscillatory neuronal circuit generating a locomotory rhythm, *Proc. Nat. Acad. Sci. USA* **73**, 3734–3738.

Glass, L. (1977a). Global analysis of nonlinear chemical kinetics, in *Modern Theoretical Chemistry*, Volume 4 (edited by B. J. Berne), Plenum, pp. 311–349.

Glass, L. (1977b). Combinatorial aspects of dynamics in biological systems, in *Statistical Mechanics and Statistical Methods in Theory and Application* (edited by U. Landman), Plenum, pp. 585–611.

Glass, L., and Pasternack, J. G. (1978a). Stable oscillations in mathematical models of biological control systems, *J. math. Biol.* **6**, 207–223.

Glass, L., and Pasternack, J. G. (1978b). Prediction of limit cycles in mathematical models of biological oscillators, *Bull. math. Biol.* **40**, 27–44.

Hastings, S. P., Tyson, J. J., and Webster, D. (1977). Existence of periodic solutions for negative feedback cellular control systems, *J. diff. Eqns.* **25**, 39–64.

Kling, U., and Szekely, G. (1968). Simulation of rhythmic nervous activities: I Function of networks with cyclic inhibitions, *Kybernetik* **5**, 89–103.

Savageau, M. (1976). *Biochemical Systems Analysis*, Addison Wesley, Reading, Pa.

Savageau, M. (1979a). Growth of complex system can be related to the properties of the underlying determinants, *Proc. Nat. Acad. Sci. USA* **76**, 5413–5417.

Savageau, M. (1979b). Allometric morphogenesis of complex systems, *Proc. Nat. Acad. Sci. USA* **76**, 6023–6025.

Thomas, R. (1979). *Kinetic Logic*, Lecture Notes in *Biomathematics* 29, Springer, Berlin.

Tyson, J. J. (1975). On the existence of oscillatory solutions in negative feedback cellular control processes, *J. math. Biol.* **1**, 311–315.

Chapter 10

Large scale clocked switching networks

In a recent statement of the biological background to his work on clocked switching networks, Kauffman (1979a) stresses the following features of the cell. Cells are rather stable systems that can be characterized as belonging to recognizable distinct types. Diverse cell types in a mature organism are distinguished by differential expression of genes and their products rather than by selective loss or amplification of genetic material during development. While there may be around 10^5 structural genes and a similar number of regulatory functions in a mammalian cell, there are only several hundred distinct cell types in a mammal. Indeed, over a range of organisms from *E. coli* to man, having widely different quantities D of DNA, the number of distinct cell types N very roughly varies as $N = aD^{\frac{1}{2}}$.

So Kauffman asks what sort of highly complex system can exhibit a relatively small number of distinct, stable, dynamic behaviour patterns. As in the regulatory processes governing metabolite synthesis, discussed in Chapter 9, so the regulatory functions of genes can be approximated by Boolean switching functions. Kauffman investigated whether systems of linked switches can be constrained, either by choice of the pattern of interactions, or by the nature of the individual Boolean functions, into a relatively small variety of stable dynamic activities. It turned out that they can, by restricting the Boolean functions to ones that give an output essentially controlled by only one of the inputs. This is readily implemented by taking all the Boolean functions to depend on two inputs, in which case all but two of the sixteen possible functions are of this kind. (These Boolean functions are set out in Table 10.1.)

The stable dynamic activities of most interest in networks of switches are cycles. So the most directly relevant biological data concern periodic phenomena, such as cell replication. Kauffman (1970) estimates the average time for a state transition in the cellular regulatory system to be one minute. Bacteria with about 200 genes typically have replication times of 10 to 100 minutes, and rarely as long as 1000 minutes. So Kauffman is interested in random switching networks of about 1000 components which exhibit, by appropriate choice of constraints on

Table 10.1 The sixteen Boolean functions of two Boolean variables $\tilde{x}$, $\tilde{y}$. In the seventh column *Fx* means forced by $\tilde{x}$, *Fy* forced by $\tilde{y}$, *Fxy* forced by both $\tilde{x}$ and $\tilde{y}$, in the sense defined in the text

$\tilde{x}$ $\tilde{y}$	0 0	0 1	1 0	1 1	Name	Forcing	Liveliness
B1	0	0	0	0	Absurdity	*Fxy*	0
B2	0	0	0	1	AND	*Fxy*	$\frac{1}{2}$
B3	0	0	1	0		*Fxy*	$\frac{1}{2}$
B4	0	1	0	0		*Fxy*	$\frac{1}{2}$
B5	1	0	0	0	NOR	*Fxy*	$\frac{1}{2}$
B6	0	0	1	1	$\tilde{x}$	*Fx*	$\frac{1}{2}$
B7	0	1	0	1	$\tilde{y}$	*Fy*	$\frac{1}{2}$
B8	1	0	0	1			1
B9	0	1	1	0	Exclusive OR		1
B10	1	0	1	0	$\bar{\tilde{y}}$	*Fy*	$\frac{1}{2}$
B11	1	1	0	0	$\bar{\tilde{x}}$	*Fx*	$\frac{1}{2}$
B12	0	1	1	1	OR	*Fxy*	$\frac{1}{2}$
B13	1	0	1	1		*Fxy*	$\frac{1}{2}$
B14	1	1	0	1		*Fxy*	$\frac{1}{2}$
B15	1	1	1	0	NAND	*Fxy*	$\frac{1}{2}$
B16	1	1	1	1	Tautology	*Fxy*	0

their structure or on the nature of the components, cycles of less than 100 time steps.

In a clocked switching network there are N elements i, each with K inputs and a single output, which also stands for the state of the element. The dynamics of each element are given by a Boolean function B_i of K Boolean variables, one for each input. There are no external inputs or unattached outputs, all outputs being connected to inputs of their own element or another. For $K > 1$, outputs are branched. The connections are assigned at random. The connections in a switching network can be described as a digraph with N nodes and KN arcs. As a simple example consider the digraph in Fig. 10.1, in which the element 1 receives inputs from its own output and that of element 3, and so on. The digraph does not completely specify the network, since the particular B_i remain to be assigned.

The network is said to be clocked because all changes of state of the elements take place simultaneously. After an interval of time τ the new outputs take effect as the inputs to give the next round of simultaneous state changes. The dynamics of the whole network can be described, as for the networks studied by Glass, by a dynamic component digraph. For example, corresponding to the digraph of Fig. 10.1 the dynamic component digraph has eight nodes, which may be labelled as

$$1 = 000,\ 2 = 100,\ 3 = 010,\ 4 = 001,\ 5 = 110,\ 6 = 101,$$
$$7 = 011,\ 8 = 111$$

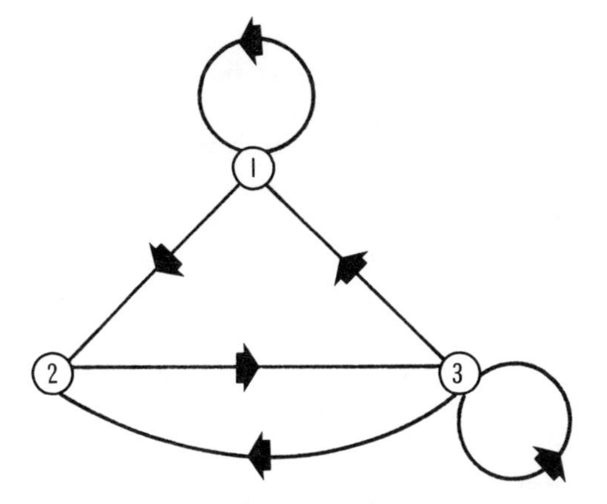

Fig. 10.1 The static digraph for a clocked switching network with $N = 3$, $K = 2$. Element 1, for example, receives inputs from its own output and from the output of element 3

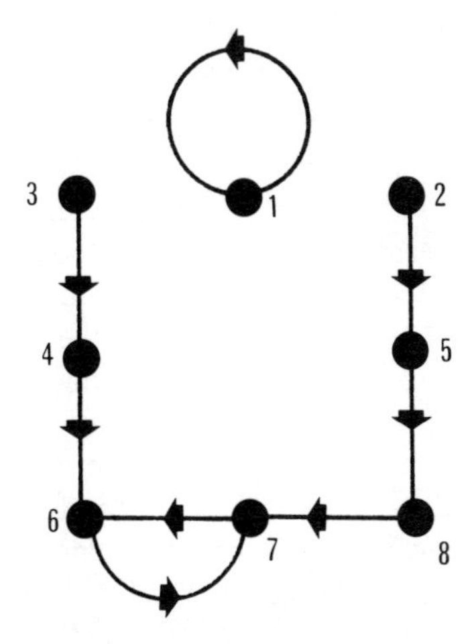

Fig. 10.2 The dynamic component digraph for the switching network shown in Fig. 10.1. States are labelled as in the text, and the transitions shown are those implied by the Boolean functions specified in the text

For example, if we specify the three Boolean functions by the notation used in Table 10.1, taking $B_1(1,3)$ as B9, $B_2(1,3)$ as B6 and $B_3(2,3)$ as B12, the dynamic component digraph is that shown in Fig. 10.2.

As in Section 9.1 but not Section 2.3, the dynamics are deterministic. Although both the static digraph and the B_i may be assigned in a random manner, once they are assigned all state changes are completely determined. Because state changes are not ordered in time, unlike those of either Section 2.3 or Section 9.1, the arcs of the dynamic component digraph are not confined to the edges of a hypercube. Because there are a finite number of states and the dynamics are deterministic, either there is a steady state (cycle of period τ) or a stable cycle. In the example of Fig. 10.2, if the network starts in state 1 it stays in that state, otherwise it finishes in a cycle between state 6 and state 7. The main question which has been studied regarding these clocked switching networks is: are these cycles mainly short compared with $N\tau$ or long?

The case $K = 1$ is a special one. Since there are no external inputs, the static digraph breaks up into parts containing $N_1, N_2, \ldots$ elements linked in loops. For each of these there may be a steady state or a cycle of period $M_r\tau$. The state of the whole network is the product of states for each of the sub-networks. Any cycle of the whole network will have a period at least the lowest common multiple of the periods $M_r\tau$. This is likely to exceed $N\tau$.

Two striking results are obtained on networks with $K > 1$. Firstly, the mean cycle length is strongly dependent on K, $K = 2$ in particular yielding short cycles. This result makes it clear that $K = 2$ networks can exhibit 'simple' behaviour. Secondly, once K and N are picked, it is the choice of B_i rather than the structure of the static digraph that determines the mean cycle length. This result gives some hope that it may be possible to account for this simple behaviour. One can seek indices to characterize the distinctions between 16, for $K = 2$, or 256, for $K = 3$, Boolean functions rather than those between a very large ($\sim K^N$) number of different digraphs.

10.1 SWITCHING NETWORKS OF WALKER AND ASHBY

I shall first describe the work of Walker and Ashby (1966) since it is simpler than that of Kauffman in the sense that all elements i are given the same Boolean function B. They take $K = 3$, giving each element a RIGHT input, a LEFT input and a SELF input, the last of these coming from the element's own output. The connections from outputs to LEFT and RIGHT inputs are made at random, and any one of the $N = 100$ elements may receive one of these inputs from its own output. A selection from the 256 possible Boolean functions is used, and for each Boolean function a set of five different networks is used. For each of these, 50 trajectories are generated by arbitrary choice of initial state, and these are followed for 500 steps. If, for any specific Boolean function, no cycle has been found in this search, a second search with 5000 steps is made.

Steady states (cycles of period τ) are found in one in three of the 128,000 runs. Since about one in six runs do not reach a cycle, there may be as many as one in two runs which would eventually reach a steady state. Some very long cycles are found. For example the Boolean function

L	R	S	
0	0	0	0
0	0	1	0
0	1	0	1
0	1	1	1
1	0	0	0
1	0	1	1
1	1	0	1
1	1	1	0

(10.1)

yields cycles of length 653τ and 2391τ. Cycles of order 100τ are quite frequently found. It is found that the distribution of cycle lengths is characteristic of the Boolean function used rather than of the specific choice of connections. So these authors set out to devise suitable indices to characterize the 256 possible Boolean functions, and to examine whether there is any noticeable correlation between the values of these indices and long cycles. The indices used are:

(1) Internal homogeneity I. This reflects the tendency of an element to give preponderantly the same value. Denoting the number of zeros in the truth table by n, I is defined as

$$I = \text{Max}(n, 8 - n) \tag{10.2}$$

and runs from 4 to 8.
(2) Hesitancy H. This reflects the tendency of the output to keep its previous value. Let the number of 1s in the truth table in rows having $S = 1$ be n_1, and let the number of 0s in the truth table in rows having $S = 0$ be n_0. Then H is defined as

$$H = n_1 + n_0 \tag{10.3}$$

and takes values from 0 to 8.
(3) Memory M. This reflects the tendency of the element to be influenced by its own output. Let the rows of the truth table be grouped in pairs which differ only in the values of S. M is the number of such pairs within which the output changes, and runs from 0 to 4. This index is not relevant to networks without the SELF input.
(4) Fluency F. This reflects the tendency of the element to be influenced by L, for fixed R, or by R, for fixed L. The value of F is found as follows. Take the four rows of the truth table with $S = 0$, match each of them with the sequence 0110 and with the sequence 1001, and record the higher score as s_0. Carry out the same

process with the other four rows and record the higher score as s_1. Then

$$F = s_0 + s_1 \tag{10.4}$$

and runs from 4 to 8.

In the example of equation (10.1) of a Boolean function associated with long cycles, I takes the minimum value 4, H takes the value 4, and M takes the value 2. The rows with $S=0$ give 0101, scoring $s_0 = 2$. The rows with $S=1$ give 0110, scoring $s_1 = 4$. So F takes the value 6.

From the graphs presented by Walker and Ashby (1966) it is apparent that a tendency to give long cycles is associated with low I, high M, and intermediate H, and that the variance of cycle length is too large to allow any clear trend with F to be identified. I shall return to the topic of the indices I, H, M, and F after discussing Kauffman's concept of a 'forcing' Boolean function.

10.2 KAUFFMAN'S NETWORKS AND THE FORCING PROPERTY

The results of Walker and Ashby (1966) have been rightly overshadowed by those of Kauffman (1969–1972) for the following reasons. By choosing to work with $K=2$, Kauffman hit on the most striking case of preponderantly short cycles. He made a more explicit and convincing case for the biological relevance of this type of model. He could relate short cycles to the presence of Boolean functions of a particular kind, his 'forcing' functions, and was able to cite examples of this kind in known biochemical control processes.

Kauffman (1972) summarizes his results on clocked switching networks with random assignment of Boolean functions B_i to the elements i, and with various K values, as follows:

(1) $K=N$, totally connected nets. State cycles are very long. They are not very abundant, the number being linearly increasing with N. They are readily perturbed to a different cycle. As in any model of a biological periodicity, it is necessary to examine the consequences of a disturbance by environmental noise; here a small disturbance is interpreted as one which makes one element give the wrong output, so throwing the system off its cyclic sequence of states. Kauffman finds that in the case $K = N$ such an error may lead to a different cycle. A recent review by Gelfand (1982) of completely random nets—in the sense that the state to follow a given one is determined, but is chosen from the 2^N possible states with equal likelihood—gives the results that the expected number of cycles is of order N, and the expected cycle length is of order $N^{-1}\,2^{N/2}$.

(2) $K=1$. Cycles can be very long, and their number rises sharply as N increases, approximately as

$$\left(\frac{2^{N/\ln N}}{2\,N/\ln N}\right)^{\frac{1}{2}\ln N} \tag{10.5}$$

They are readily perturbed to about N other cycles.

(3) $K = 2$. Cycles are of period about $N^{\frac{1}{2}}\tau$, and there are about $N^{\frac{1}{2}}$ of them. They are very stable, about 90% of perturbations having no long-term effect, and such perturbations that do lead to a new cycle yield one of a rather small set of cycles.
(4) $K = 3$. State cycles are long, unless the B_i are chosen in a very special manner, in which case the results can be brought to resemble those for $K = 2$. To understand this choice, and why it is essentially already made for us if we take $K = 2$, it is necessary to appreciate the significance of Kauffman's concept of a forcing Boolean function. (In later papers, such as Kauffman (1979b), he uses the term 'canalising' instead of 'forcing'.)

A forcing Boolean function of K (>1) Boolean variables is one that takes a particular value when one variable takes one of its two possible values, irrespective of the values taken by the other $K - 1$ variables. In Table 10.1, for example, the Boolean function B8 can take the value 0 or 1 for $\tilde{x} = 0$, for $\tilde{x} = 1$, for $\tilde{y} = 0$, or for $\tilde{y} = 1$. We have to know both $\tilde{x}$ and $\tilde{y}$ to predict the output. So B8, and in a similar manner B9, is not forcing. B6, which is simply $\tilde{x}$, is a forcing function. It is forced by $\tilde{x} = 0$ to the forced value 0, and by $\tilde{x} = 1$ to the forced value 1. B2 = AND is a forcing function, being forced by $\tilde{x} = 0$ or by $\tilde{y} = 0$ to the same forced value 0. (In general, when a Boolean function is forced by more than one variable, the forced value is the same in each case.) Even excluding the cases B1 and B16, which are unlikely to feature in control processes, in a $K = 2$ network there is likely to be a 6:1 predominance of forcing B_i.

It is possible to define a forcing index, which could, for example, run from 0, for a function which is not forcing, to $2K$ for the analogues of B1 and B16. Aleksander (1973), in a paper amplifying some of Kauffman's analysis, uses an index which he calls 'liveliness' which is complementary to a forcing index. The liveliness of an element is the probability that a randomly chosen input will, when its value changes, yield a change in the value of the output, given that all the other inputs are held constant at randomly chosen values. The last column of Table 10.1 gives the liveliness of the Boolean functions B1 to B16.

In contrast to the case $K = 2$, forcing functions are relatively rare for $K = 3$. Table 10.2 shows the distribution of the forcing property among the 256 Boolean functions of three variables. The variables are interpreted as the LEFT, RIGHT and SELF inputs in the model of Walker and Ashby (1966), and classified according to their indices I, H, M, and F. Low I, which these authors found to be associated with long cycles, correlates with a low probability that an element is forcing. For $I = 7$ to 8, all the B are forcing. For $I = 6$, six out of seven of the B are forcing; for $I = 5$, two out of five; and for $I = 4$, only one in ten. High M, which Walker and Ashby also found to be associated with long cycles, is also correlated with low probability that an element is forcing. For $M = 0$, nine cases out of eleven are forcing, while for $M = 1$, 2, 3, and 4 this ratio falls successively to five out of seven, two out of seven, one out of four, and one out of ten. Thus in the model of Walker and Ashby (1966) there is a correlation between a low incidence of forcing Boolean functions and the presence of long cycles.

Before presenting Kauffman's arguments to the effect that this correlation is to

Table 10.2 The numbers of forcing (F) and non-forcing (N) Boolean functions in the model of Walker and Ashby, divided according to the values of the indices I (internal homogeneity), H (hesitancy), M (memory), and F (fluency)

	I	8	7	6	5	4
H	0					F
	1				$4F$	
	2			$6F$		$16N$
	3		$4F$		$7F$ $17N$	
	4	F		$12F$ $4N$		$4F$ $32N$
	5		$4F$		$8F$ $16N$	
	6			$6F$		$16N$
	7				$4F$	
	8					F
M	0	F		$4F$		$4F$ $2N$
	1		$8F$		$15F$ $9N$	
	2			$20F$ $4N$		$48N$
	3				$8F$ $24N$	
	4					$2F$ $14N$
F	4	F		$8F$		$6F$ $12N$
	5		$8F$		$23F$ $17N$	
	6			$16F$ $4N$		$48N$
	7				$16N$	
	8					$4N$

be expected, it is necessary to introduce some more definitions. Kauffman defines the relationship between two elements, 'i forces j' in this way:

> If i has a forcing B_i and the output of i feeds one input, say the pth input, of j, and if B_j is a forcing function on its pth input when that input takes the forced value of B_i, then i is said to force j. If, by this definition, i forces j and j forces k, then i is said to force k (indirectly).

For example, in a network with $K = 2$, let B_i and B_j both be the OR function B12, and let the output of i be connected to one input of j. B12 is forced by either $x = 1$ or $y = 1$, and its forced value is 1. So in this case i forces j. Had B_i been B2, which has the forced value 0, with B_j still B12, then i would not force j. Kauffman next defines the forcing structure of an element as all those elements which either force this element or are forced by it, directly or indirectly. If an element belongs to its own forcing structure the network contains a forcing cycle.

If we consider a digraph with a node for each element of the network and an arc (i, j) whenever the output of element i is connected to an input of element j, then this definition of forcing structure corresponds to the presence of a sub-digraph, which may be called the **forcing digraph**, which includes only those elements

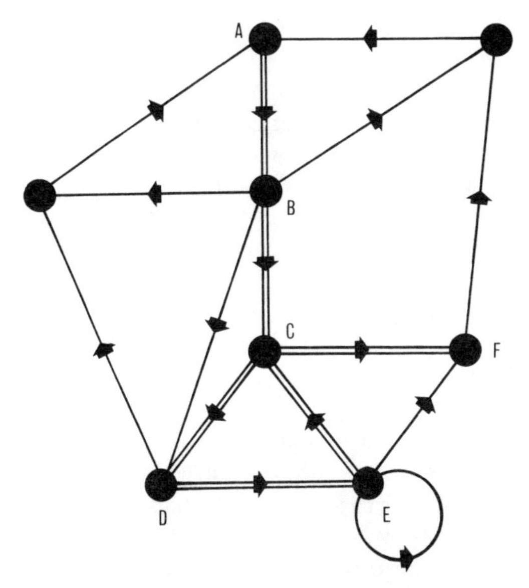

Fig. 10.3 The forcing structure (double lines) of element C in a switching network with $K = 2$

which belong to forcing structures, and only those arcs corresponding to the connections which do the forcing. If the network has separate forcing structures, the forcing digraph is disconnected. If a forcing structure contains a forcing cycle, the forcing digraph contains a closed path. Fig. 10.3 illustrates this.

The network of Fig. 10.1 is not a forcing cycle if we adopt the Boolean functions used to obtain the dynamics that are summarized in Fig. 10.2, because $B_1(1,3)$ is B9 which is not forcing. The other Boolean functions are forcing; $B_2(1,3)$ is forced by input $\tilde{x}$ (from 1) since it is B6 $= x$, and $B_3(2,3)$ is forced by input $\tilde{y} = 1$ (from 2), the forced value being 1. To make the network a forcing cycle it suffices to take $B_1(1,3) =$ B7, which is forced by $\tilde{y} = 1$ (from 3) to take the value 1. Either B12 or B14 would also have this effect.

When embedded within a network, forcing cycles are important because they ultimately arrive at fixed outputs for all their elements, each taking a forced value. Consider the possibilities at the time of a state change, for a single element in a forcing structure. Either it receives the forced value on a forcing input, in which case it must pass on its own forced value which forces the next element in the structure, or it does not. If it does not then its output is influenced by the non-forcing inputs, and in a large random network it has an equal chance of passing on its forced value or of not doing so. Forced values can enter at any element, and do not leave the structure. When they appear in a cyclic part of the structure they are permanently preserved. So forcing cycles are quite quickly 'frozen out' of the dynamics of the network, and any cyclic behaviour of state changes must come from other parts of the network. In the forcing structure illustrated in Fig. 10.3, for example, the part leading into the cycle (elements A, B) can suffer changes of output, but the cycle (elements C, D, E) and its 'tail' F cannot, once forced values are established in the cycle.

Further, if forcing structures predominate, the state dynamics of the whole network must ultimately be the dynamics of isolated 'live' (that is, non-forcing) sub-networks. In fact, it is possible to show that there is a critical value for the proportion of elements belonging to forcing structures; if the proportion rises beyond this value live sub-networks are of small dimensions and isolated from one another. This situation is rather analogous to one occurring in percolation and allied physical phenomena. For example, let us imagine a square sheet of metal foil, with electrical current flowing into one corner and out at the opposite one. Let small holes be punched out of the foil one after another at random points. The resistance of the foil slowly increases. Then as the proportion of the area of the foil removed reaches a critical value, the sheet changes quite rapidly from a continuous one with a lot of holes to a broken one with no path for the current.

In this context a basic graph result was established by Erdos and Rényi (1960). They work with a set of V vertices, and add successive edges at random. This means that if m edges are present each of the remaining $\frac{1}{2}V(V-1)$ pairs of vertices is equally likely to be joined by an edge at the following step. At $m=\frac{1}{2}V$ a collection of small graphs is replaced by a graph which spans about $V^{\frac{2}{3}}$ vertices.

Forcing structures predominate in the case $K=2$, and become relatively less abundant as the proportion of forcing B_i falls with increasing K. Kauffman (1972) gives upper limits, in terms of N and K, for such quantities as the expected number of forcing inputs, and the expected number of connections belonging to forcing structures. These limits are proportional to N, and fall off strongly as K increases. For example, the expected number R of forcing connections in a network of N elements is subject to

$$R < N F(K)$$

where (10.6)

$$F(2) = 4, \; F(3) \simeq 1, \; F(4) \simeq 2 \times 10^{-3}$$

Thus the computed results of Kauffman and of Walker and Ashby are consistent with the idea that it is the forcing behaviour of the individual switches that leads to the relatively simple and stable dynamics of the network. Kauffman (1979b) discusses the evidence that particular genetic control mechanisms display the appropriate choice of Boolean function. He cites in particular the lac operon and the arabinose operon in *E. coli* and two regulatory processes in bacteriophage lambda, involving respectively the right operator and the left operator.

Newman and Rice (1971) emphasize that in models with continuous functional dependence on the variables, the analogue of a forcing structure is one in which all the functions are monotonic. They argue that monotonicity forces the variables into a relatively small region of their potential range, and that this is a necessity for regulatory biochemical interactions.

It is possible to look in more detail at the statistics of cycle length in Kauffman's networks, as in those of Walker and Ashby. Kauffman (1969) presents some results for $K = 2$ and $N = 400$, giving the relative probabilities $P(n)$ for cycles of length $n\tau$. There is a pronounced, and unexplained, bias towards even n; once this is smoothed out, the trend of $P(n)$ is as n^{-1}. It may be relevant that Aleksander (1973) finds that the number of rings of 'live' elements falls off as the reciprocal of the number of elements in the ring. Sherlock (1979) has made a detailed study of this aspect of the clocked switching networks. He emphasizes the state dynamics rather than the nature of connections in the network. He shows how a state transition matrix can in principle be reduced to an $N_c \times N_c$ matrix involving only the states which take part in cyclic activity. On the assumption that the statistics of cycles of length $n\tau$ can be approximated in terms of the statistics of random partitions of N_c, he arrives at the result

$$P(n) = an^{-1} \tag{10.7}$$

in agreement with Kauffman's empirical trend. He points out that this result, together with Kauffman's result that for $K = 2$ the mean cycle length is $N^{\frac{1}{2}}\tau$, leads to the conclusion that the total number of states taking part in cycles is approximately

$$N_c = N \tag{10.8}$$

References

Aleksander, I. (1973). Random logic nets: stability and adaptation, *Int. J. Man Machine Stud.* **5**, 115–131.

Erdos, P., and Rényi, A. (1960). On the evolution of random graphs, *Magyar Tud. Akad. Kutato Inst. Kozl.* **5**, 17–61.

Gelfand, A. E. (1982). A behavioural summary for completely random nets, *Bull. math. Biol.* **44**, 309–320.

Kauffman, S. A. (1969). Metabolic stability and epigenesis in randomly constructed genetic nets, *J. theor. Biol.* **22**, 437–467.

Kauffman, S. A. (1970). Behaviour of randomly constructed genetic nets: binary element nets, in *Towards a Theoretical Biology* 3: *Drafts* (edited by C. H. Waddington), Edinburgh University Press, pp. 18–33.

Kauffman, S. A. (1971). Gene regulation networks: a theory for their global structure and behaviour, *Curr. Top. Dev. Biol.* **6**, 145–182.

Kauffman, S. A. (1972). The organisation of cellular genetic control systems, in *Lectures on Mathematics in the Life Sciences* (edited by J. D. Cowan), Amer. Math. Soc., pp. 61–116.

Kauffman, S. A. (1979a). The molar behaviour of cells in development, in *Kinetic Logic* (edited by R. Thomas), Lecture Notes in *Biomathematics* 29, Springer, Berlin, pp. 13–29.

Kauffman, S. A. (1979b). Assessing the possible regulatory structures and dynamics of the metazoan genome, in *Kinetic Logic* (edited by R. Thomas), Lecture Notes in *Biomathematics* 29, Springer, Berlin, pp. 30–60.

Newman, S. A., and Rice, S. A. (1971). Model for constraint and control in biochemical networks, *Proc. Nat. Acad. Sci.* **68**, 92–96.

Sherlock, R. A. (1979). Analysis of the behavior of Kauffman binary networks I, II, *Bull. math. Biol.* **41**, 687–705 and 707–724.

Walker, C. C., and Ashby, W. R. (1966). On temporal characteristics of behaviour in certain complex systems, *Kybernetik* **3**, 100–108.

Part III

Branching Structures: Description, Biophysics, and Simulation

'The total amount of air that enters the trachea is equal to that in the number of stages generated from its branches, like a plant in which each year the total estimated size of its branches when added together equals the size of the trunk.'

Leonardo da Vinci, *Windsor MS*

'From every leaf in all the countless crowd at the tree's summit, one slender fibre . . . descends through shoot, through spray, through branch and through stem.'

J. Ruskin, *Modern Painters*

'They must not be too small, or the work of driving blood through them will be too great; they must not be too large, or they will hold more blood than is needed—and blood is a costly thing.'

d'Arcy W. Thompson, *Growth and Form*

Chapter 11

Branching structures in biology: topology and geometry

Attempts to analyse biological branching structures have a long history. In order to introduce the concepts discussed in the next few chapters I shall trace this history no further back than 1950 and pick out a few contributions, on dendrites, the lung, and the arterial system. Sholl (1953) introduced a method of statistical analysis of the dendritic tree of a nerve cell, which was extended by later authors and criticized by ten Hoopen and Reuver (1970). The limitations of Sholl's analysis indicate the advantages to be gained by separating the topological from the geometrical aspects even of the relatively simple branching structures studied by these authors. The topological aspects are those retained if the branching structure is treated as a tree in the mathematical sense defined in Section 1.2. The reader should recall the definitions of tree, binary tree, root and leaf given in that section. The branching structures to be described in the following chapters are treated, so far as their topology is concerned, as binary rooted trees, with the root adjacent to only one edge. The geometrical aspects usually regarded as most significant are the lengths and diameters of segments, and the angles at which segments meet at bifurcations. Geometrical properties may be considered either globally—for example, are the finer vessels also the shorter vessels?—or locally, for example, is there a relationship between the branching angles and the ratios of diameters at individual bifurcations?

In describing rooted binary trees of this kind it is usual to adopt some scheme by which edges can be grouped into different orders. Several such schemes will be described in the next chapter, but for the moment it will suffice to point out two different types of scheme, centrifugal and centripetal. In the first, the unique edge adjacent to the root is nominated as the sole edge of order 1, the two which stem from it at the first bifurcation are assigned to order 2, and so on. In the second, all the edges which are adjacent to a leaf are assigned to order 1, two of them come together to form an edge of order 2, and so on. Results described by ten Hoopen and Reuver (1970) indicate a certain advantage in a centripetal scheme. The book by Weibel (1963), which made a substantial contribution to the quantitative morphology of the lung, also focuses attention on the consequences of choosing a particular centrifugal scheme.

Possibly the most striking feature of the arterial branching system is the substantial increase in the total cross-sectional area of the segments of a given order, as one proceeds centrifugally from the aorta. The reviews by Green (1950) and by Iberall (1967) are largely concerned with this aspect, and with qualitative ideas on the significance of the number of orders from aorta to capillaries.

11.1 DENDRITIC TREES

Sholl (1953) studied the dendrites of 30 neurons in the visual and motor regions of the cerebral cortex of the cat. He noted that the dendrites of stellate cells (Fig. 1.1) and the basal dendrites of pyramidal cells (Fig. 11.1) are distributed in a tolerably symmetrical manner around the cell body. So he proposed to describe the spatial distribution of the dendrites in terms of the number $c(r)$ of dendrites that cut a spherical surface of radius r centred on the cell body. Sholl typically used ten equally spaced concentric surfaces, as well as recording the number $c(0)$ of dendrites leaving the cell body. Neglecting the fact that $c(0)$ is not zero, he fitted his values of $c(r)$ with the expression

$$c(r) = 4\pi r^2 a \exp(-kr) \tag{11.1}$$

This allowed an estimate of the mean radius of the dendritic distribution as

$$\bar{r} = 24\pi \, a \, k^{-4}$$

Subsequent authors, for example Eayrs and Goodhead (1959), pointed out that $c(r)$ can be counted in a different way, as

$$c(r) = c(0) + B(r) - E(r) \tag{11.2}$$

where $B(r)$ is the number of bifurcations within a spherical surface of radius r, and $E(r)$ is the number of dendrite endings within this surface. Both with direct counting and when equation (11.2) is used, it is assumed that only an insignificant proportion of the dendrites double back. As pointed out by ten Hoopen and Reuver (1970), equation (11.2) reveals a limitation of Sholl's method. The same set of values of $c(r)$ can be obtained, for a given $c(0)$, by a variety of sets of values of $B(r)$ and $E(r)$, so long as the difference between these two is unchanged at each r. For example, the same $c(r)$ could be recorded in a branching structure with many bifurcations giving rise to short segments terminating mostly within the next spherical surface, and in one with fewer bifurcations into longer segments, extending across several further surfaces. This is illustrated in Fig. 11.2. Sholl's method, in fact, does not distinguish between the topological aspect of frequency of bifurcation and the geometrical aspect of mean segment length.

Ramon Moliner (1967) pointed out that the most typical branching pattern in dendritic trees involves a marked increase in the length of segments as one goes outward from the cell body. When ten Hoopen and Reuver (1970) reanalysed much of the earlier data, in particular those of Peters and Bademan (1963) on

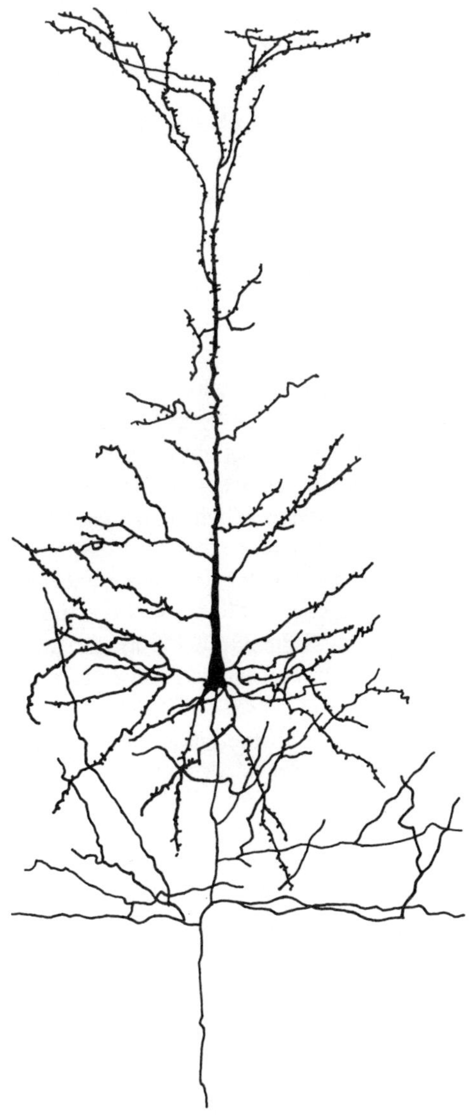

Fig. 11.1 A pyramidal cell from the brain of a mouse. The basal dendrites are the spiky branches emerging from the central mass. The axon (below) has some smooth branches. The apical branches are at the top of the figure. Redrawn by M. F. MacDonald from an illustration in Ramon y Cajal, *Histologie du Système Nerveux*, Paris, 1909

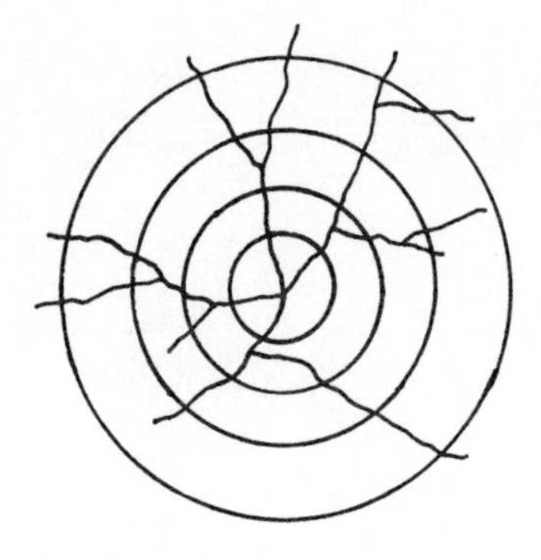

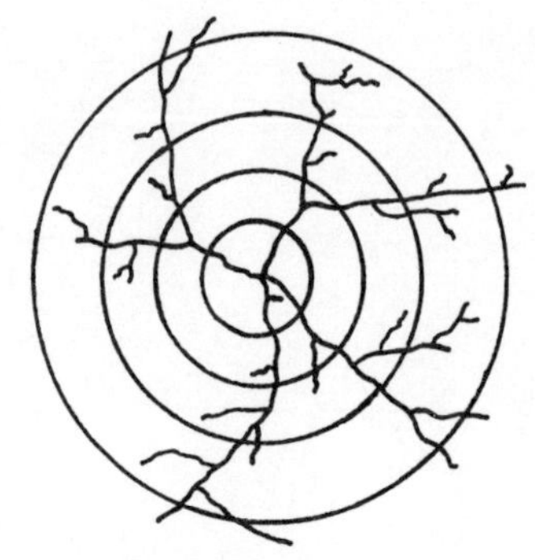

Fig. 11.2 Illustrating how, in Sholl's method of describing the geometry of the dendritic tree, the same sequence of values $c(r)$—4, 7, 9, 7—can occur for two rather different branching structures

stellate cells in the cortex of the guinea pig, they found that terminal segments are typically about twice as long as segments which end in a bifurcation. They adopted a centrifugal ordering scheme, in which all dendrite segments leaving the cell body are assigned to order 1. (In the standard description adopted here we would have to treat the dendrites as belonging to a number of trees, each having its root at the cell body.) They found that the probability that a segment bifurcates falls off with order n as $p_n \propto p_1^n$. So they noted both geometrical and topological regularities, in these relatively simple branching structures, which have typically only four orders, and in which $c(r)$ peaks at about twenty. It can be argued that the geometrical result, indicating that terminal segments are unusually long, is in favour of the policy of grouping these as the segments of order 1, in a centripetal scheme.

11.2 LUNG AIRWAYS

Weibel (1963) describes the morphology of the lung airways in the following terms. Starting from the trachea the number of airways multiplies by branching, and the wall structure of the branches changes as one goes towards finer vessels. This corresponds to the change from purely conductive airways to ones in which gas transfer takes place in the small cuplike openings called alveoli. 'Bronchi' have walls of a passive character, with a thin muscle layer and rings or plates of cartilage. 'Bronchioles' have walls consisting mostly of active smooth muscle. 'Respiratory bronchioles' have the same wall structure as bronchioles, but interrupted by occasional clusters of alveoli. 'Alveolar ducts' have walls covered by alveoli in a mesh of muscle fibres. 'Alveolar sacs' are the terminal branches, alveolar ducts with their ends closed off. It is possible that the organization of branching in the purely conductive zone differs from that in the zone containing alveoli.

Most of Weibel's quantitative discussion is in terms of a symmetric model, in which each branch, unless it is an alveolar sac, divides symmetrically into two

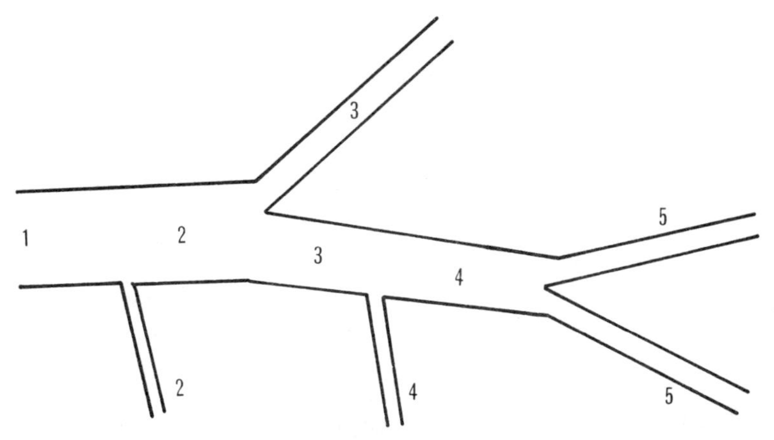

Fig. 11.3 The consequences of Weibel ordering when a major vessel gives off successive minor vessels

branches of the next (higher) order. (He takes the trachea to have order 0.) He finds that this gives 24 successive generations or orders. Because the branching of the lung is, in fact, rather asymmetric, the method leads to a considerable spread of the dimensions of the branches taken to belong to a given order. It also assigns the alveolar sacs to a wide range of orders. Later authors consider these to be disadvantages of Weibel's ordering scheme, and have also drawn attention to another disadvantage. This is illustrated in Fig. 11.3, which represents a major branch giving off a number of minor branches. It is somewhat artificial to require that the order of each successive segment of the main branch should increase by one. The more asymmetric the branching structure, the more critical becomes the choice of an appropriate ordering method. We shall see that in a great variety of branching structures it is possible to specify the asymmetry by a single index, the Horton branching ratio.

In terms of his symmetric model Weibel presented results for the average diameter $\bar{d}_m$ of branches of order m. Excluding the trachea and the first two bronchi—orders 0 and 1—and extending through about ten orders of bronchi and bronchioles, he fitted these numbers with

$$\bar{d}_m = 2^{-m/3} \tag{11.3}$$

To obtain values for this fit he had to make quite strong assumptions about the statistics of the dimensions of unobserved small vessels. Equation (11.3) implies an increase in total cross-sectional area of branches of one order, as the order increases. Sample respiratory bronchioles and alveolar ducts are wider than is suggested by extrapolation of equation (11.3). So Weibel's results indicate, in addition to the qualitative distinctions between four types of vessels according to wall structure, a division into three zones—the first few large vessels, the rest of the bronchi and the bronchioles, for which equation (11.3) holds, and the finer vessels in which gas transfer takes place.

11.3 ARTERIES

In the data on the arterial branching system compiled by Green (1950) and by Iberall (1967) similar features emerge. Green, basing his conclusions mainly on a much earlier investigation of the mesenteric arterial system of a dog, introduced a classification into nine levels—aorta; large arteries; main, secondary and tertiary branches; terminal arteries; terminal branches; arterioles; capillaries. The number at each level increases by a factor of ten on average, but by no means regularly. The mean diameter falls by a factor $\frac{1}{2}$ or $\frac{1}{3}$ from each level to the next. The geometry of the aorta and large arteries is dictated largely by the gross anatomy of the animal, while statistical regularities apply mostly to the finer vessels. As with the lung, Green's data indicate a great increase in the total cross-section at the highest levels.

Iberall (1967) made an extensive compilation of data on arterial branching, including much not available to Green. Data on arteries with diameter greater than 1 mm were taken from the work of Patel *et al.* (1961, 1963) on a 23 kg dog. By suitable scaling the data of Mall (1905), which extend to much finer vessels, could be matched to those of Patel *et al.* However, the most precise data on finer vessels used by Iberall were those of Suwa *et al.* (1963) on human arteries. For diameters above 1 mm, there is no change of total cross-section with branching order. Between 1 mm and 30 μm, there is a substantial increase; in the work of Suwa *et al.* this is expressed in terms of a relationship between the three diameters at a bifurcation

$$d_0^{2.7} = d_1^{2.7} + d_2^{2.7} \tag{11.4}$$

We shall encounter power law relations of this kind in several contexts later. Small vessels display a more rapid increase in total cross-section.

Iberall emphasizes that Suwa *et al.* find a continuous distribution of vessel diameters, so that division into 'levels' or orders cannot be made solely from the measurement of diameters. He suggests that the idea of a level should be related to the presence of a main branch, giving off a number of minor branches (which belong to levels one or more higher) and bifurcating finally into two branches of the next higher level. He interprets the data as suggesting about $N = 11$ levels, with $R = 8$ as the approximate ratio of the numbers of branches in successive levels. (With this notation, Green has $N = 9$ and $R = 10$.) The number of capillaries is about 8^{11} or 10^9. These are very crude estimates, and it is not clear that the concept of a level has any real significance. The need for a systematic topological ordering scheme, quite independent of vessel diameters, is obvious.

Both Weibel's results, assuming a symmetric scheme, and the approach adopted by Iberall draw attention to a potentially useful question to raise once data on an asymmetric branching structure have been organized according to an acceptable ordering scheme. Does the number of branches belonging to order m depend on m in a regular manner, for example is there a fixed ratio R of numbers in successive orders? And if so, what can we learn from this?

In conclusion I wish to emphasize that the policy of treating an arterial branching system, or even some compact part of it, as a tree has distinct limitations. Cross-linking of branches, technically known as anastomosis, is very common. As an example I shall cite the paper by May (1967) on the arteries in the head of the sheep. He draws attention to the great variety of anastomoses, dividing them into three main categories. There are anastomoses between two or more branches supplying the same tissue, for example the transverse facial artery system in the masseter muscle. There are anastomoses between branches of two or more arterial systems, the most extensive being between the maxillar labial, infraorbital and malar arteries in the nose and upper lip. There are anastomoses across the median plane, in particular between homonymous arteries on the two sides. The anastomosing vessels have diameters 1 mm or less.

Before any further discussion of the statistics of branching patterns, and of the length and diameters of branches, in these and other biological examples, I shall digress in the next chapter to describe some of the work of geographers who were concerned with ordering networks of streams. The debate on the choice of ordering method arose earlier in that discipline than in biology, and the consensus that emerged among the geographers has influenced the choice of one particular method—Strahler ordering—by the majority of biologists engaged in this kind of analysis. Again, regularity in numbers of branches in successive orders was emphasized by geographers, and this influenced the biologists to seek such regularity.

References

Eayrs, J. T., and Goodhead, B. (1959). Postnatal development of the cerebral cortex of the rat, *J. Anat.* **93**, 385–402.

Green, H. (1950). Circulatory system: physical principles, in *Medical Physics*, Vol. 2, (edited by O. Glasser), Yearbook Publishers, New York, pp. 228–251.

Iberall, A. S. (1967). Anatomy and steady flow characteristics of the arterial system, with an introduction to its pulsatile characteristics, *Math. Biosci.* **1**, 375–395.

Mall, F. (1905). A study of the structural unit of the liver, *Amer. J. Anat.* **5**, 227–340.

May, N. D. S. (1967). Arterial anastomoses in the head and neck of the sheep, *J. Anat.* **101**, 381–387.

Patel, D., Mallos, N., and Fry, D. (1961). Aortic mechanics in the living dog, *J. appl. Physiol.* **16**, 293–299.

Patel, D., de Freitas, F., Greenfield, J. G., and Fry, D. (1963). Relationship of radius to pressure along the aorta in living dogs, *J. appl. Physiol.* **18**, 1111–1117.

Peters, H. G., and Bademan, H. (1963). The form and growth of stellate cells in the cortex of the guinea pig, *J. Anat.* **97**, 111–117.

Ramon Moliner, E. (1962). An attempt at classifying nerve cells on the basis of their dendritic patterns, *J. comp. Neurol.* **119**, 211–227.

Sholl, D. A. (1953). Dendritic organisation in the neurons of the visual and motor cortices of the cat, *J. Anat.* **87**, 387–406.

Suwa, N., Niwa, T., Fukusawa, H., and Sasaki, Y. (1963). Estimation of intravascular blood pressure by mathematical analysis of arterial casts, *Tohoku. J. exp. Med.* **79**, 168–198.

ten Hoopen, M., and Reuver, H. A. (1970). Probabilistic analysis of dendritic branching patterns of cortical neurons, *Kybernetik* **6**, 176–188.

Weibel, E. R. (1963). *Morphometry of the Human Lung*, Springer, Berlin.

Chapter 12

Law and order in trees

Geographers have proposed various ways of ordering the branches in a stream network. It is natural to turn to their work for guidance in how to deal with analogous features of other branching structures, since they have the advantage of copious ready-made data. A geographer, while he may be concerned with the effect of map scale on the apparent complexity of a stream network, does not have to make his own maps. Biologists with an interest in quantitative aspects of biological branching structures have to go through very laborious processes to obtain good data, and in consequence finish up with rather small samples. Thus Horsfield *et al.* (1976) cite data on six canine lungs, while Shreve (1966) cites data on over 200 stream networks.

12.1 ORDERING METHODS

An early ordering method proposed by Gravelius (1974) requires a decision at each bifurcation as to which branch is the main stream. Here, as in what follows, it is assumed that the simultaneous confluence of three or more streams is unlikely. Also anastomoses, such as can happen towards the mouth of a river, are ignored. The method involves starting at the river mouth—the root of the binary rooted tree—and tracing out a path upstream, taking at each junction the stream making the smaller angle with the direction in which one has entered the junction. It is plausible to assume that this is usually the stream contributing the greater flow into the junction, and the choice by angle can be made by reference to the map alone. The process is continued until the presumed main stream has been traced to its source, and is then repeated for each tributary of the main stream, and so on until the network has been exhausted. This is illustrated in Fig. 12.1.

Contemporary interest in stream ordering stems largely from the work of Horton (1945), who drew attention to a number of empirical regularities, usually now known as Horton's laws. His ordering method, however, is most succinctly described as a variant of one proposed later by Strahler (1952). Start with the leaves of the rooted tree, that is with those vertices, excluding the root, which are

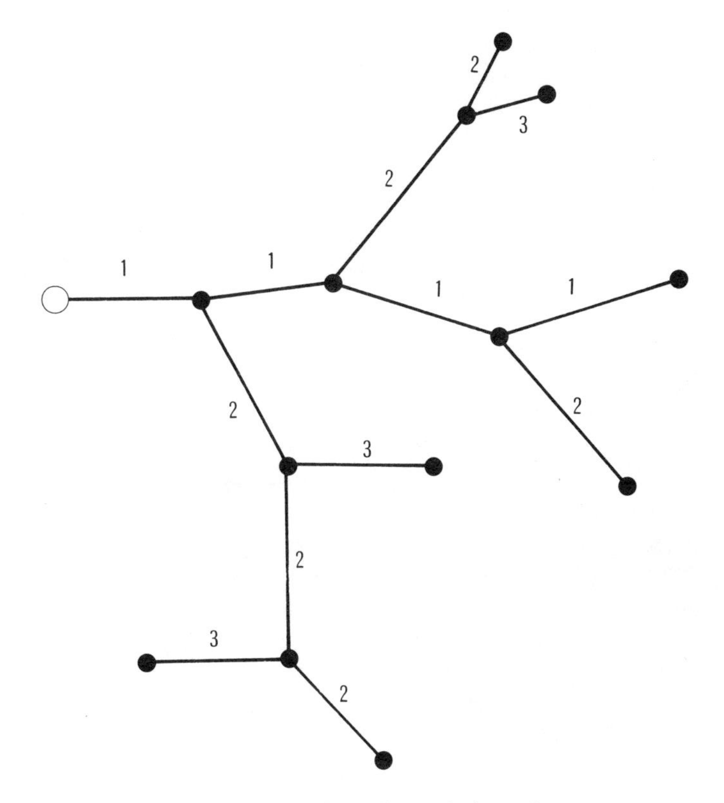

Fig. 12.1 The Gravelius ordering scheme

contiguous to only one edge. All edges that are contiguous to a leaf are branches of order 1. These are streams with no tributary. Where two branches of order m come together, the third edge contiguous to that vertex belongs to a branch of order $m + 1$. Where a branch of order m meets a branch of order n, where n exceeds m, the third edge at the vertex is a continuation of the branch of order n. So we introduce here a new technical term in graph theory, **branch**, as a set of consecutive edges that arise in such an ordering scheme. There are no arbitrary decisions in Strahler ordering, and it is purely topological. Strahler ordering is illustrated in Fig. 12.2.

Horton ordering requires one further step after a preliminary Strahler ordering. This is to retrace all branches outward, at each bifurcation of the type $(m \to m-1, m-1)$ selecting the main stream, for example by the rule of the smaller angle as in the Gravelius method. This is assigned to be a continuation of the branch of order m, so that the bifurcation is now $(m \to m, m-1)$. The effect of this is illustrated in Fig. 12.3. Like Gravelius' method, this mixes geometry with topology, but it is not simply a reversed Gravelius order.

Shreve (1966) introduced a method that avoids the need for branches, as distinct from edges, and which qualitatively reflects the cumulative increase in

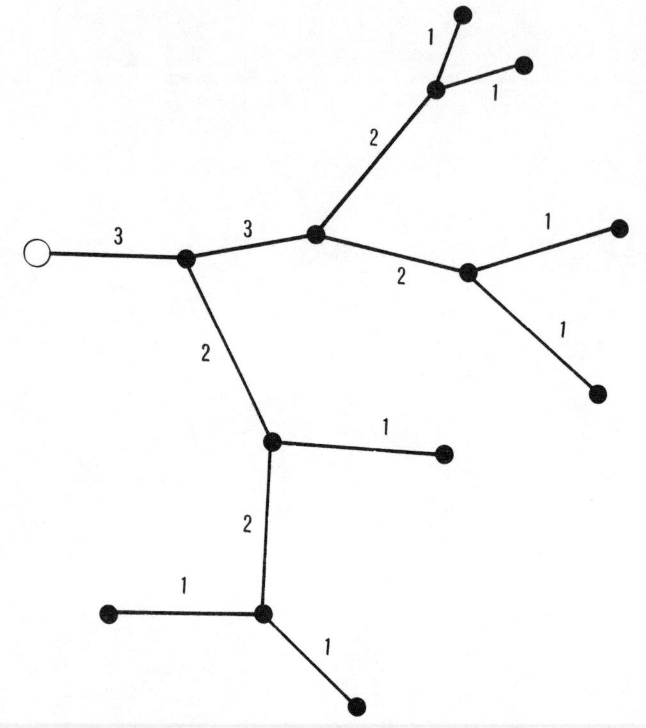

Fig. 12.2 The Strahler ordering scheme

flow as tributaries come together. Start with edges assigned to order 1 as in the Strahler method, but instead of Strahler's two rules apply a single rule, that when an edge of order m and an edge of order n come together, the third edge at that vertex has order $m + n$. This is illustrated in Fig. 12.4. Further variants are described by Haggett and Chorley (1969). The four so far mentioned serve to illustrate the choices available:

centrifugal (Gravelius) or centripetal (all others);
cumulative (Shreve) or non-cumulative (all others);
purely topological (Strahler, Shreve) or partly geometrical (Gravelius, Horton);
with edges only (Shreve) or with branches (all others).

The figures can be used to introduce a further useful term. We shall say, following Prosser (1982), that a branch or an edge **subtends** all those branches of Strahler order 1 (terminal branches) that can be reached from it by successive bifurcation. Thus in Fig. 12.2 the third order branch subtends seven, and the lower second-order branch subtends three terminal branches. The Shreve order (Fig. 12.4) of an edge is identical with the number of terminal branches subtended by that edge.

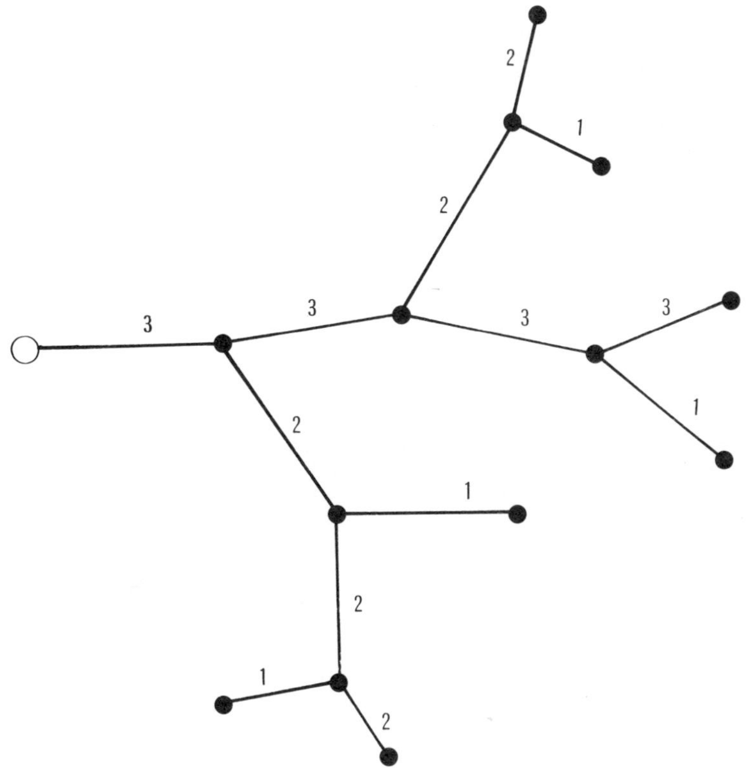

Fig. 12.3 The Horton ordering scheme

While I am mainly concerned in this chapter with the work of geographers, it is appropriate to describe here the method of Weibel (1963) and another, described by Horsfield *et al.* (1971) and also applied to lung airways, so that they may be compared with the four methods just described. Weibel's method is centrifugal, topological, non-cumulative and with edges only. Take the edge contiguous to the root to have order 0. At each vertex reached along an edge of order m, take the other two edges as of order $m + 1$. This is shown in Fig. 12.5. In the long sequence of papers on lung airways by Horsfield and his collaborators, the ordering method originally adopted is related to that given by Shreve (1966), except that when edges of order m and order n come together at a vertex, the third edge is assigned not to $m + n$ but to order one greater than the greater of m and n, or to $m + 1$ if $m = n$. So this ordering scheme is centripetal, topological and has edges only. In papers from 1973 onwards, Horsfield and his collaborators have adopted Strahler ordering, which facilitates comparison of their results with those of other authors. However, the scheme just described is the basis for a method, described by Horsfield *et al.* (1971), for retaining asymmetric features in models otherwise

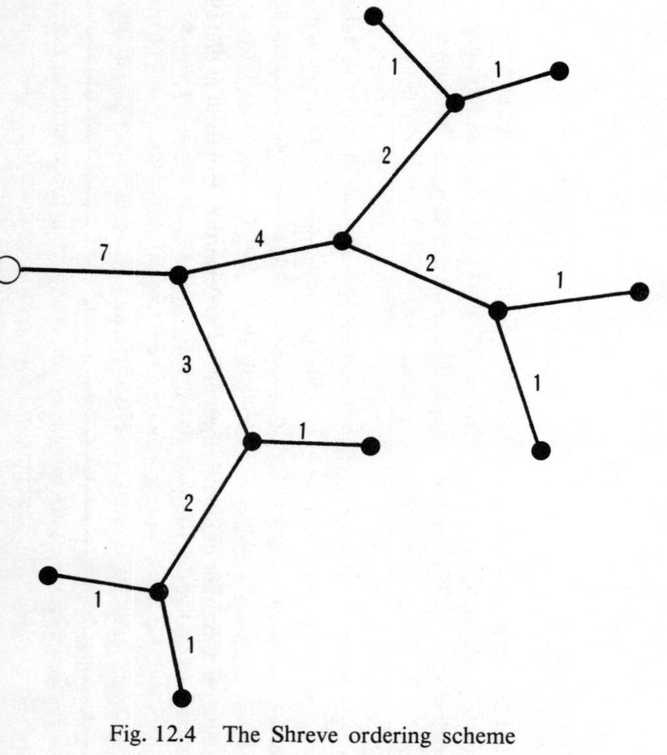

Fig. 12.4 The Shreve ordering scheme

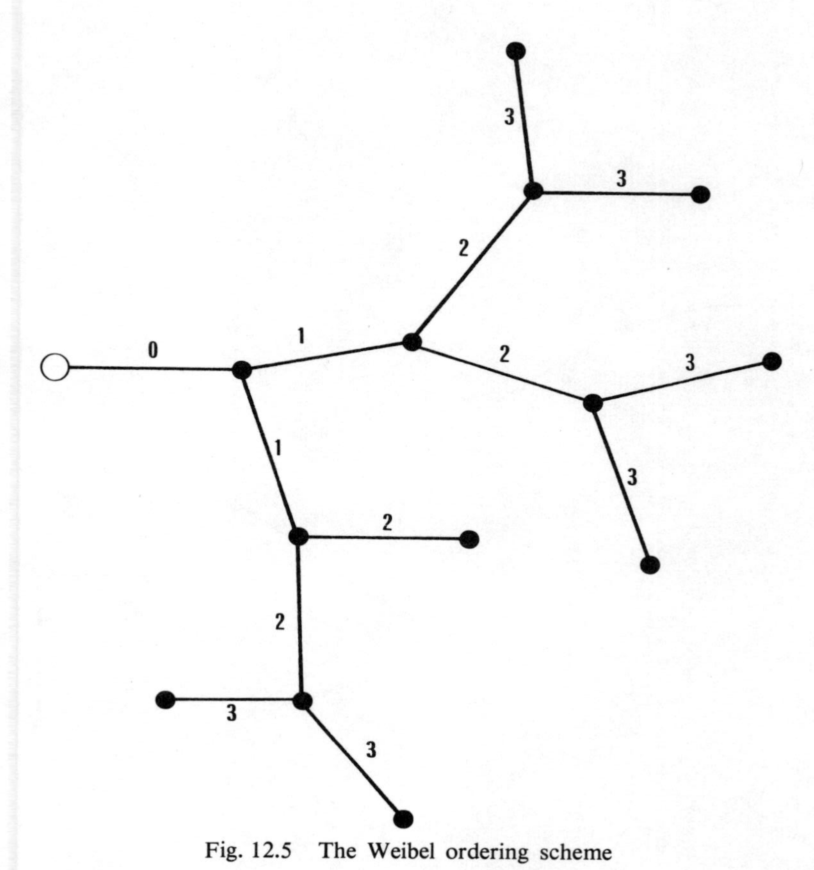

Fig. 12.5 The Weibel ordering scheme

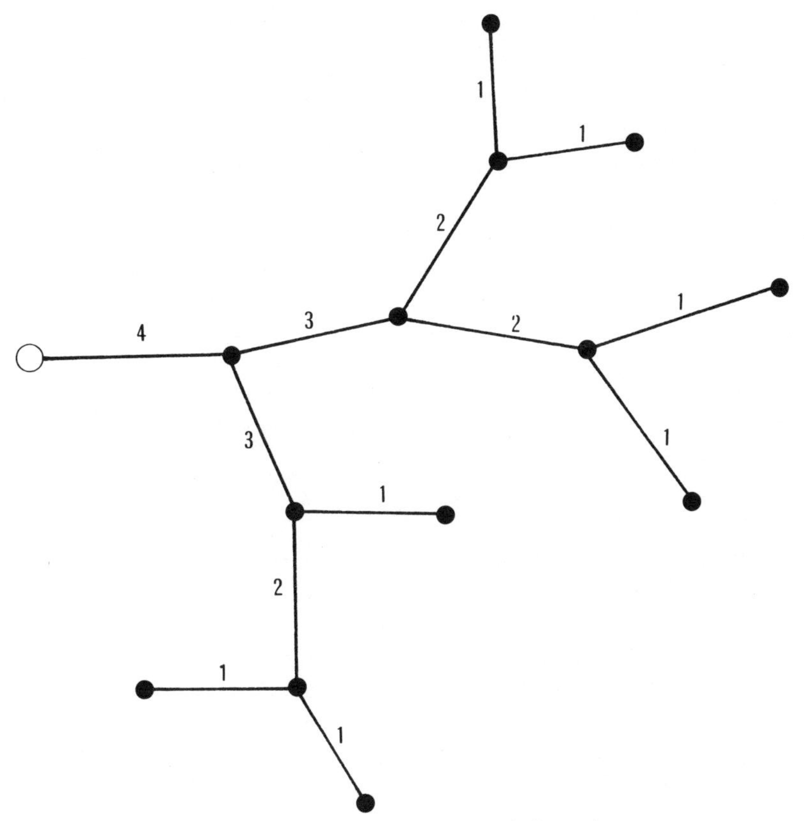

Fig. 12.6 The Horsfield ordering scheme

as simple as the symmetric model used by Weibel (1963). This I shall discuss in Section 14.3 and Section 16.3. Horsfield ordering is shown in Fig. 12.6.

12.2 HORTON'S LAW OF THE BRANCHING RATIO

It is apparent from these figures that even in this simple example the different methods give different assessments of the number of orders. This effect becomes more pronounced when there are many branches, such as that illustrated in Fig. 11.3, giving off a number of lateral branches of lower order. Horsfield *et al.* (1982) give a striking example from their analysis of a cast of the lung of a dog. This extends over ten Strahler orders but over 43 Horsfield orders.

Before going on to discuss Horton's law of the branching ratio, some algebraic results are needed. The first relates the number H_m of branches of order m in the Horton scheme to the numbers N_m and N_{m+1} of branches of order m and of order $m + 1$ in the Strahler scheme

$$H_m = N_m - N_{m+1} \tag{12.1}$$

The second relates the total number of edges E in a binary rooted tree, to the number of edges of Strahler order 1

$$E = 2N_1 - 1 \tag{12.2}$$

The proof of this is by induction. Assume that the result holds for $N_1 = N$. At one of the leaves add two new edges, so that the original vertex is no longer a leaf, but two new leaves have been added. So E is increased by 2, N_1 by 1, and the result holds for $N_1 = N + 1$. (An alternative way of raising N_1 by 1 is to insert a new vertex in the middle of an edge, with an appended new edge carrying a leaf. This again increases E by 2.) Now the result holds for a tree consisting of a single edge joining the root to a single leaf, in which $E = N_1 = 1$. So the result holds for any N_1.

The third result is an inequality, following directly from the definition of Strahler ordering. Since no branch of order $m + 1$ arises except by the confluence of two branches of order m

$$N_m \geqslant 2N_{m+1} \tag{12.3}$$

The difference of these two quantities may be called the number of excess branches of order m. These are lateral branches off branches of any order higher than m, not necessarily from branches of order $m + 1$. Because their allocation to the higher order branches is arbitrary, we do not have a result for the number of edges which make up all branches of a given order. There is, however, one useful extension of equation (12.2). Let us make a truncated tree by removing all branches of order less than m. The number of edges in this tree—which is not equal to the number of edges of order m or greater in the original tree—is

$$E_m = 2N_m - 1 \tag{12.4}$$

Horton's empirical branching law is that stream networks conform approximately to the result

$$\frac{H_m}{H_{m+1}} = R_H \tag{12.5}$$

where the best value of the constant R_H is found by plotting log H_m against m. Later authors, for example Shreve (1966), who surveys data on 246 stream networks, find that the law is obeyed rather more precisely if Strahler orders are used, so that equation (12.5) is replaced by

$$\frac{N_m}{N_{m+1}} = R \tag{12.6}$$

This is one reason why geographers prefer Strahler ordering. A simple consequence of equation (12.6) is that on the average a branch of Strahler order m subtends R^{m-1} branches of order 1.

Conflicting statements appear in the geological literature concerning the relationship between equations (12.5) and (12.6). Let us write down the successive numbers of branches in each order, taking note of equation (12.1), and assuming that equations (12.5) and (12.6) both hold precisely with integer R and R_H. In the highest order we have in each case one branch, in the next lowest

$$R_H = R - 1$$

in the next lowest

$$R_H^2 = R^2 - R = R(R - 1)$$

in the next lowest

$$R_H^3 = R^3 - R^2 = R^2(R - 1)$$

and so on. Strictly speaking, as emphasized by Shreve (1966), these results are inconsistent. If all are correct we must have $R = R_H$ *and* $R - 1 = R_H$. However, if we set aside the second highest order, it is possible for the numbers of branches over a succession of orders to obey both equation (12.5) and (12.6) with $R = R_H$.

Another result, again strictly only applicable for integer R, but leading to a convenient approximate rule, is that the total number of branches is

$$\begin{aligned} N_B &= \sum_{m=1}^{M} N_m \\ &= 1 + R + R^2 + \cdots R^{M-1} \\ &= \frac{R^M - 1}{R - 1} \end{aligned}$$

So the fraction of branches which are terminal ones (Strahler order 1) is

$$\frac{N_1}{N_B} = \frac{R^{M-1}(R-1)}{R^M - 1} \simeq 1 - R^{-1} \tag{12.7}$$

12.3 RANDOM BINARY TREES

Geographers were for a while concerned with attempts to find a geomorphological basis for Horton's branching law, and for the fact that the values of R typically lie in the range three to four. These efforts were somewhat undermined by the results of Shreve (1966) on random binary trees in a plane. The significance of the term **tree in a plane** is that the left or right orientation of a lateral branch of order m from one of some higher order is meaningful. This must be the case for a stream network, but not for an entirely abstract tree, or presumably for one describing branching in three-dimensional space. Shreve investigated whether Horton's law would hold for randomly constructed trees in a plane, and if so what value of R would be most likely. He made use of a simple combinatorial argument, based on equation (12.4) and the inequality (12.3), to obtain the number $N(N_1, \ldots, N_{M-1})$ of random binary trees in the plane which have maximum Strahler order M, and prescribed numbers N_m in each order from $m = 1$ to $m = M - 1$. (There is only one branch of maximum order.) From the N_m branches of order m, a number $2N_{m+1}$ must come together to make the N_{m+1} branches of the next higher order. The $N_m - 2N_{m+1}$ excess branches are attached at random to the $2N_{m+1} - 1$ edges of the truncated tree having only the branches of order $m + 1$ and higher. These lateral branches can be allocated in F_m ways, where

$$F_m = \begin{bmatrix} (2N_{m+1} - 1) + (N_m - 2N_{m+1}) - 1 \\ N_m - 2N_{m+1} \end{bmatrix}$$

$$= \begin{bmatrix} N_m - 2 \\ N_m - 2N_{m+1} \end{bmatrix} = \frac{(N_m - 2)!}{(N_m - 2N_{m+1})!\,(2N_{m+1} - 2)!} \tag{12.8}$$

The orientation of each lateral branch gives another factor

$$2^{(Nm-2Nm+1)} \tag{12.9}$$

so that the result is

$$N(N_1, \ldots, N_{M-1}) = \prod_{m=1}^{M-1} F_m 2^{(Nm-2Nm+1)} \tag{12.10}$$

Later Shreve (1967) proved that if a random binary tree is treated as a sample sub-tree of an infinite random binary tree, Horton's law is obeyed with $R = 4$. It is easy to see that for large N_1 the factors F_m tend to give $R = 4$ for the lower orders of branching. The binomial coefficient $\binom{a}{b}$ has its maximum value when $b = a/2$ or $b = (a \pm 1)/2$. In equation (12.8), when N_m is large, this corresponds to

$$4N_{m+1} = N_m \tag{12.11}$$

Of course the successive factors are not independent, and the additional powers of two complicate matters. As an example of the results obtained for fairly small trees in a plane, Werrity (1972) quotes the case $N_1 = 50$. The possible sets of N_m range from

$$50,\ 25,\ 12,\ 6,\ 3,\ 1\colon\ R = 2$$

to

$$50,\ 1\colon\ R = 50$$

Sixty per cent of the trees constructed using equation (12.10) belong to the number sequences shown in Table 12.1, obeying Horton's law quite closely, with R around 4.

If we omit the powers of two, as we should in dealing with trees which are not in a plane, a certain bias towards lower R is to be expected. For low R means fewer

Table 12.1 The percentage probability of the occurrence of the ten most abundant sets of Strahler order numbers N_2 and N_3 when $N_1 = 50$. $N_4 = 1$ in each case, this being the highest order present, although order 6 is possible for $N_1 = 50$, which exceeds 2^5

N_2	N_3	%
12	3	9.7
13	3	9.0
11	3	7.1
13	4	6.3
14	3	5.7
14	4	5.3
12	4	4.8
12	2	4.2
11	2	4.1
10	3	3.5

excess branches in each order, and so less opportunity for the right or left choice. Indeed, for $N_1 = 50$ the most abundant number sequences, shown in Table 12.2, have values of R around 3. We shall see that in the lung observed R values are around 3. Since the lung is a branching system with N_1 measured in millions, the expected R on the basis of random branching is presumably nearer the asymptotic value 4 than the value 3 found for small N_1.

The complexity of a branching structure conforming with Horton's branching law can be characterized by quoting R and M. Any branching structure in which the rules for branching are different in different orders, so that R cannot be defined, will have an additional dimension of complexity.

Although I describe them as random trees, the binary trees constructed by Shreve have a certain regularity, in that the rules do not change from order to order. So the significance of Horton's branching law is that it indicates a degree of topological self-similarity in the branching structure. This has encouraged several authors to investigate regularities corresponding to geometrical self-similarity, such as

$$\frac{\bar{d}_m}{\bar{d}_{m-1}} = R_d \tag{12.12}$$

Table 12.2 The most abundant sets of Strahler order numbers N_2, N_3, and N_4 when $N_1 = 50$, and no distinction is made between branches to the right and to the left. The first is three times as common as the last

N_2	N_3	N_4
16	5	2
15	5	2
16	6	2
14	5	2
17	6	2
15	6	2
14	4	1
14	4	2
16	5	1
16	4	1
15	5	1
15	4	1
15	4	2

$$\frac{\bar{l}_m}{\bar{l}_{m-1}} = R_l \tag{12.13}$$

where $\bar{d}_m$ and $\bar{l}_m$ denote the mean diameter and mean length of a branch of order m. The constants R_d and R_l are defined in this way throughout the following chapters, with the higher order in the numerator, so that R, R_d and R_l are all greater than one. Empirically branches are almost always shorter and thinner as one goes towards order 1, although there can be local exceptions, as in the terminal dendrites of the stellate cells mentioned in the previous chapter.

In the next chapter I shall present data from a wide variety of contexts, all resulting from analysis in terms of the Strahler scheme. For many only R is available, but there are a selection of R_d values for trees, fossil plants and bronchial trees, and some R_l values for dendrites and bronchial trees. I shall examine the relationships between these three ratios, and compare them with certain predicted values to be discussed in more detail in later chapters.

References

Gravelius, H. (1914). *Flusskunde*, Goschen, Berlin.

Haggett, P., and Chorley, R. J. (1969). *Network Models in Geography*, Methuen, London.

Horsfield, K., Dart, G., Olson, D. E., Filley, G. F., and Cumming, G. (1971). Models of the human bronchial tree, *J. appl. Physiol.* **31**, 207–217.

Horsfield, K., Relea, F. G., and Cumming, G. (1976). Diameter, length and branching ratios in the bronchial tree, *Resp. Physiol.* **26**, 351–356.

Horsfield, K., Kemp, W., and Phillips, S. (1982). An asymmetrical model of the airways of the dog lung, *J. appl. Physiol.* **52**, 21–26.

Horton, R. E. (1945). Erosional development of streams and their drainage basins: hydrophysical approach to quantitative morphology, *Bull. Geol. Soc. Amer.* **56**, 275–370.

Prosser, J. I. (1982). Growth of fungi, in *Microbial Population Dynamics* (edited by M. J. Bazin), CRC Press, Boca Raton, pp. 125–166.

Shreve, R. L. (1966). Statistical law of stream numbers, *J. Geol.* **74**, 17–37.

Shreve, R. L. (1967). Infinite topologically random channel networks, *J. Geol.* **75**, 178–186.

Strahler, A. N. (1952). Hypsometric (area-altitude) analysis of erosional topography, *Bull. Geol. Soc. Amer.* **63**, 1117–1142.

Weibel, E. R. (1963). *Morphometry of the Human Lung*, Springer, Berlin.

Werrity, A. (1972). The topology of stream networks, in *Spatial Analysis in Geomorphology* (edited by R. J. Chorley), Methuen, London, pp. 167–196.

Chapter 13

Branching ratios, branch length ratios, and branch diameter ratios with Strahler ordering

It is difficult to compare topological features, or global geometrical features, of different branching structures, unless they have been ordered in the same way. Fortunately most recent studies employ the Strahler scheme as described in the previous chapter, and illustrated in Fig. 12.2. In this chapter I shall present a table of data on a variety of biological branching structures, all ordered according to the Strahler scheme. The quantities presented in Table 13.1 are the maximum order (which may be for the whole structure or some part of it) and the ratios R, R_d, and R_l defined in equations (12.6), (12.11), and (12.12). The three ratios are compared in Figs 13.2–13.5, which also display the predictions of some models to be examined later.

13.1 THE RATIOS R, R_d, AND R_l

The two major compilations of R values for living trees are those of Oohata and Shidei (1971) and of Whitney (1976). The Japanese authors find that the R values for deciduous trees, running from 2.93 to 3.97, do not overlap to any great extent with those for evergreen trees, which run from 3.64 to 6.53. However, the values determined by Whitney for deciduous trees run from 3.08 to 8.62. Steingraeber *et al.* (1979) emphasize the wide variation possible in the values of R within one species, when individual trees grow in very different conditions. The sugar maple, growing under a forest canopy, disposes its leaves in an umbrella form, with only a small departure from symmetric branching, the mean R being about three. When it grows in the open the leaves are disposed in roughly conical or cylindrical form, with many side branches from the main stem, and the mean R is about seven.

McMahon and Kronauer (1976), who analyse only a few cases, develop a detailed mechanical model of tree form, which will be discussed in Section 16.1.

Table 13.1 The ratios R, R_d and R_l for Strahler ordered trees

Details	Samples	Maximum order	R	R_d	R_l	
TREES Oohata & Shidei (1971)						
Ilex pendunculosa	2	5	3.81, 4.21			Four broad-leaved evergreen species, with mean $R = 4.05$
Pasania edulis	2	5	3.64, 3.71			
Vaccinium bracteatum	1	5	4.10			
Cinnamonum camphora	2	6	4.34, 4.55			
Alnus firma	1	4	3.89			Five broad-leaved deciduous species, with mean $R = 3.01$
Clethra barberinus	1	6	3.55			
Castanea crenata	2	5	2.93, 3.17			
Lyonia ovalifolia	2	6	3.10, 3.18			
Rhododendron dilatatum	3	7	2.95, 3.42 3.97			
Metasequoia glyptostroloides	1	6	2.99			A deciduous conifer
Abies firma	1	4	3.89			Four evergreen conifer species, with mean $R = 5.53$
Pinus thunbergii	5	5	3.97, 4.65, 5.30, 5.51, 6.53			
Picea jezoensis	1	6	5.26			
Pinus densifolia	4	5	4.52, 5.56, 5.78, 6.19			
TREES Barker *et al.* (1973)						
Malus	1	5	4.35	1.90		
Betula	1	6	4.00	1.94		

Table 13.1 *continued*

Details	Samples	Maximum order	R	R_d	R_l	
TREES Holland (1969)						
Eucalyptus incrassata	4	6, 7	2.7, 3.2, 3.2, 3.4			
Eucalyptus dumosa	10	5–7	3.0, 3.1, 3.2, 3.2, 3.4, 3.4, 3.5, 3.8, 3.9, 4.0			
Eucalyptus oleosa	5	4, 5	3.3, 3.5, 4.0, 4.0, 4.0			
TREES Whitney (1977)						
*Populus tremuloides**	1	5	8.62			Ten deciduous simple broad-leaved species
*Prunus pennsylvanicus**	1		5.54			*compare McMahon and Kronauer value
Lindodendron tulipifera	1		5.25			†R values given for whole tree and for crown
Betula populifera†	1		4.21/5.07			‡see Steingraeber *et al.*
Betula alleghanensis	2		3.46, 3.43			
Quercus velutina	1		4.90			
Acer rubrum	1		6.41			
Acer saccharum‡	1	6	4.35			
Fagus grandifolia	1		3.50			
Cornus florida	1		3.75			
Fraxinus americana	7	3–5	3.64, 3.70, 4.16, 3.76, 4.03, 4.08, 5.89			Two deciduous compound broad-leaved species (the mean value given for this group is not consistent with the individual values)

Rhus typhina	2	4	3.10, 3.08		
Kalmia labifolia	1		2.99		Two evergreen broad-leaved species
Rhododendron maximum	1		3.20		
Tsuga canadensis	1		6.20		Two evergreen conifer species
*Pinus strobus**	2		4.40, 4.15		
TREES McMahon & Kronauer (1976)					
*Quercus rubra**	1	6+	3.83	1.56	These authors use 'segment' for Strahler branch
Quercus alba	2	6	4.11, 4.37	1.84, 1.87	
Populus tremuloides†	1	6	4.22	1.86	*large trees, in which only the six highest orders, down to branches of diameter 1 cm, were counted
Prunus pennsylvanica†	1	4	5.18	2.05	†compare with data of Whitney
Pinus strobus†	1	5	4.44	2.04	
TREES Steingraeber *et al.* (1979)					
Acer saccharum (forest grown)	20	4–7	mean 3.19 ± 0.33 range 2.78–3.80		These authors studied saplings to survey the range of R values encountered in different conditions of growth
Acer saccharum (in open ground)	20	3–5	mean 7.05 range 3.78–12.25		
DIVARICATES Tomlinson (1978)					
Aristotelia fruticosa	5		3.56, 3.71, 3.36, 3.04, 2.99		The author remarks that preliminary data suggest that Horton's second law does not apply, that is to say that one cannot obtain R_l. He uses 'segment' for 'branch'.
Coprosma cf. propinqua	1		3.59		
Corokia cotoneaster	4		3.51, 3.25, 3.68, 3.06		
Discaria toumatou	1		3.12		
Hymenanthera alpina	1		3.58		
Melicope simplex	1		4.03		

Table 13.1 *continued*

Details	Samples	Maximum order	R	R_d	R_l
Melicytus micranthus	1		3.08		
Muehlenbeckia astoni	4		3.34, 3.34, 2.59, 2.88		
Muehlenbeckia complexa	3		4.18, 5.74, 4.17		
Myrtus obcordata	2		4.14, 4.74		
Nothofagus cliffortioides	2		4.62, 4.36		
Nothopanax anomalum	1		3.57		
Olearia virgata	3		3.69, 2.86, 3.24		
Pennantia corymbosa	3		3.99, 4.13, 4.73		
Plagianthus betulinus	7		4.07, 3.36 5.74, 4.79, 4.47, 2.60, 4.57		
Plagianthus divaricatus	2		3.58, 4.51		
Sophora prostrata	6		4.61, 3.93, 3.90, 4.83, 3.57, 4.15		
Teucridium parvifolium	3		3.50, 5.39, 3.88		

FOSSIL PLANTS Niklas (1978)					
*Rhyniophyta**	14		2.0–2.14	1.00–1.07	*Summary of data. In the detailed set given I have omitted most of the Rhyniophytes, the Zosterophyllophytes, two Trimerophytes and one Progymnosperm because the examples studied had very few orders
*Trimerophyta**	11		2.0–3.5	1.2–2.05	
*Zosterophyllophyta**	5		4.3–4.6	3.0–3.14	
*Progymnospermia**	12		3.14–3.88	1.19–1.82	
Eogaspesia gracilis	1	8†	2.09	1.04	
Hicklingia edwardii	1	9†	2.02	1.02	
Psilophyton princeps	1	7†	2.65	1.35	
Psilophyton dapsile	1	7†	2.32	1.80	
Psilophyton microspinosum	1	7†	2.38	1.62	
Trimerophyton robustus	1	5†	3.26	1.38	†Niklas does not give the number of orders, but he does give the total number of segments. So his values of *R* and equation (12.2) allow these estimates of the maximum order
Trimerophyton sp.	1	6†	3.32	1.45	
Perlica varia	1	7†	3.20	1.85	
Perlica quadriforia	1	11†	2.00	1.95	
Perlica sp.	1	9†	2.03	1.75	
Oocampsa calleta	1	6†	3.52	2.01	
Chaleusa cirrosa	1	6†	3.88	1.30	
Tetraxylopteris schmidtii	1	6†	3.26	1.46	
Tetraxylopteris ashlandicum	1	6†	3.22	1.52	
Triloboxon sp.	1	5†	3.26	1.21	
Archaeopteris fissilis	1	6†	3.37	1.20	
Archaeopteris obtusa	1	5†	3.38	1.23	
Archaeopteris maculata	1	7†	3.36	1.19	
Aneurophyton sp.	4	5,6†	3.14, 3.14, 3.16, 3.18	1.39, 1.72 1.77, 1.53	

Table 13.1 *continued*

Details	Samples	Maximum order	R	R_d	R_l	
DENDRITES Hollingworth & Berry (1975)						
Purkinje cells of rat	6	6	mean 3.31		mean 1.51	
Purkinje cells of rat	10	7	mean 2.88		mean 1.29	
ARTERIES Singhal *et al.* (1973) and Horsfield (1978)						
Human pulmonary arteries, diameter over 0.8 mm	1	~5	2.99			Three orders of major vessels are excluded because they do not display regular branching
Intermediate orders	1	~5	3.10			Two different techniques give different estimates of R for the finest vessels
Finest arteries in this system	1	~5	3.15, 3.60			
ARTERIES Singhal *et al.*, quoted by Woldenberg (1972)						
Pulmonary arteries of human left lung	1		3.59			
Pulmonary arteries of human right lung	1		3.24			
VEINS Horsfield & Gordon (1981)						
Human pulmonary veins	3	~9	3.16	1.47	1.50	
BRONCHI Raabe *et al.* (1976)						
Human	2		2.51, 2.74	1.35, 1.45	1.33, 1.46	
Dog	2		3.58, 3.04	1.53, 1.51	1.69, 1.53	
Rat	1		3.31	1.53	1.82	
Hamster	1		3.26	1.50	1.92	

BRONCHI Horsfield & Thurlbeck (1981)						
Sheep	2		3.53, 3.57	1.67, 1.71	1.51, 1.39	
BRONCHI Horsfield *et al.* (1976)						
Human	1	15	2.81	1.43	1.40	
BRONCHI Horsfield, quoted by Woldenberg (1972)						
Human left lung	1		2.76	1.40		
Human right lung	1		2.82	1.40		
BRONCHI Horsfield *et al.* (1976)						
Dog	6	15	3.26	1.50	1.60	The third example is a one-week-old dog, the others are mature dogs (R_l plots appreciably less regular than R_d plots)
			3.38	1.56	1.62	*proximal zone (6 or 7 orders)
			3.06	1.52	1.71	†intermediate zone (4 orders)
			3.61* 3.05†	1.63	1.68	
			3.47* 3.43†	1.56	1.66	
			3.45* 3.27†	1.48	1.56	

A major compilation of data on fossil plants (Niklas 1978) also relates to work done with that model. Niklas emphasizes the extreme difference between the R values of the early Rhyniophytes (2.00 to 2.14) and the later Zosterophyllophytes (4.3 to 4.6) and Progymnosperms (3.14 to 3.88), regarded as evolutionary descendants of the Rhyniophytes. McMahon and Kronauer (1976), Niklas (1978), and also Barker *et al.* (1973) all present values of R_d as well as of R.

Tomlinson (1978) presents R values for divaricating shrubs, a group of plants conspicuously present in New Zealand. These have been defined as having

> much branched often stiff and wiry stems which are pressed close together or even interlaced.

Tomlinson points out that 'much branched' is not an accurate description, since the R values lie in the lower half of the range covered by the data of Oohata and Shidei.

In passing, a possible source of confusion in these botanical papers may be noted. Tomlinson (1978), like McMahon and Kronauer (1976), uses the term 'segment' for a branch, *not* for an edge.

The main compilations of airways data for mammalian lungs, employing this ordering scheme, are those of Raabe *et al.* (1976) and Phalen *et al.* (1978) and those to be found in a long series of papers by Horsfield and his collaborators (Horsfield and Cumming 1968a,b, Horsfield *et al.* 1971, Parker *et al.* 1971, Horsfield and Cumming 1976, Horsfield *et al.* 1976, Horsfield 1977, Woldehirst and Horsfield 1978, Thurlbeck and Horsfield 1980, Horsfield and Thurlbeck 1981, Horsfield *et al.* 1982, Bowes *et al.* 1982). In the first few of these, in which the data are presented in the greatest detail, these authors devised their own ordering scheme, described in Section 12.1. They became convinced of the advantages of the Strahler scheme and subsequently re-analysed their earlier data in terms of this scheme. Table 13.1 gives both R and R_d values for all these lung examples.

Table 13.1 also contains a few values of R and R_l for the dendrites of Purkinje cells in the cerebellum of the rat (Hollingworth and Berry 1975) and a few values of R and R_d for pulmonary arteries (Singhal *et al.* 1973, Horsfield 1978). The branching pattern of the pulmonary arteries is known to follow that of the bronchi but with some additional branches not correlated with those of the bronchial tree. So it is natural to find R somewhat larger for human pulmonary arteries than for the human bronchial tree. Horsfield and Gordon (1981) give an estimate of R for human pulmonary veins.

Some data for biliary ducts and veins in the liver are quoted by Woldenburg (1972) from a thesis by Williamson (1967). Some work has been published on branching patterns in the mycelium of fungi, but it is difficult to compare it with the cases tabulated here. Gull (1975) employed Horton ordering in the analysis of the branching of young colonies of *Thamnidium elegans*, having usually three and sometimes four orders. Ho (1978) analysed branching patterns at the edge of mature colonies of various Phycomycetes. Horton's law did not hold. The data of Cheetham *et al.* (1980) on colonial animals are too sparse, and the branching structures too simple, to yield a significant value of R. In other colonial animals, such as *Podocoryne carnea* (Braverman and Schrandt 1967) there is extensive anastomosis, and this kind of analysis is not appropriate. For fascinating non-quantitative detail on the branching structures, many of them not binary, found in tropical trees, the reader is referred to the book of Hallé *et al.* (1978).

13.2 POWER LAWS

In the following three chapters I shall discuss some theoretical investigations bearing on the relationships between the three ratios R, R_d, and R_l. Here I shall just quote some of the results, which are in the form of power laws

$$R_d = R^{\alpha}$$

or

$$R_l = R^{\beta}$$

In the pipe model of Shinozaki *et al.* (1964a) for botanical trees, the cross-sectional area of a branch is proportional to the number of leaves supported by it, directly or indirectly. This can be interpreted as proportionality to the number of branches of Strahler order 1 subtended by the branch in question, which implies

$$R_d = R^{\frac{1}{2}} \tag{13.1}$$

As evidence for the pipe model, Shinozaki *et al.* (1964b) present data on the total leaf weight W_L of 45 birch trees aged 7 to 45 years, displayed against the square of the diameter d_B at a point on the trunk just below the first branch which bears leaves. This is fitted quite well by a linear dependence on d_B^2. They also present a similar plot of total branch weight W_B against d_B^2, from 32 larch trees of ages 11 to 41 years. Again a linear dependence is apparent. On the other hand Murray (1927) presents evidence, from measurements on nine deciduous species, that the total weight W_d distal to a cut with diameter d varies proportionally to $d^{5/2}$. Consider cuts on either side of a bifurcation, close enough in for the two distal weights to be very similar. Then in the notation of Fig. 13.1

$$W_d \propto d^{\gamma}$$

implies

$$d_0^{\gamma} = d_1^{\gamma} + d_2^{\gamma} \tag{13.2}$$

So Murray's data and the data on Shinozaki *et al.* indicate respectively

$$d_0^{5/2} = d_1^{5/2} + d_2^{5/2}$$

and

$$d_0^2 = d_1^2 + d_2^2$$

with the implication that Murray's data suggest a lower increase of R_d with R than in equation (13.1).

Arguments discussed in Chapter 15, treating the lung airways or the arterial

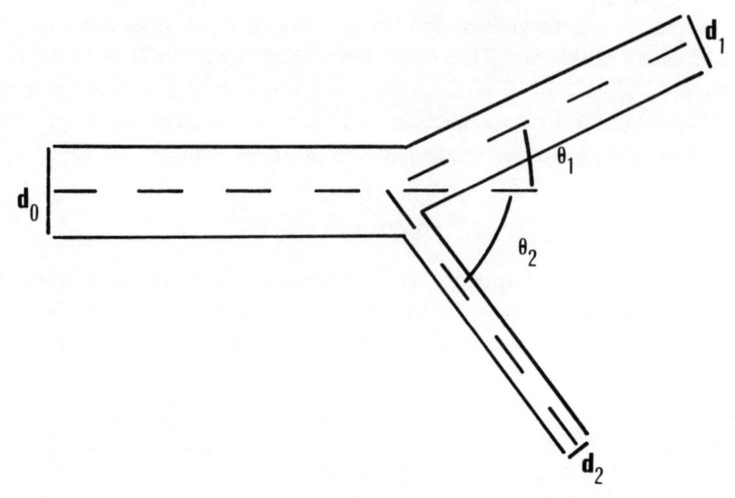

Fig. 13.1 A parent vessel, of diameter d_0, bifurcating into two daughter vessels of diameters d_1 and d_2. The centre lines of the two daughter vessels make different angles θ_1 and θ_2 with the centre line of the parent vessel

system as an efficient mechanism for passing gas or blood respectively to where it is needed, suggest comparing the ratios R_d and R in the relevant data with

$$R_d = R^{\frac{1}{3}} \tag{13.3}$$

Suwa *et al.* (1963), assuming that blood flow should correlate with some power of vessel diameter, looked for a relation between powers of diameters at a bifurcation. Their data are summarized in Table 13.2 and can be expressed by

$$d_0^{2.7} = d_1^{2.7} + d_2^{2.7}$$

Now in the case $R = 2$, equation (13.3) implies

$$d_0^3 = d_1^3 + d_2^3$$

So the data of Suwa *et al.* (1963) suggest a more rapid increase of R_d than given by equation (13.3).

We shall see in Chapter 14 that a symmetric branching structure ($R = 2$), treated as a means of uniformly filling a volume with the end points of branches of Strahler order 1, requires that R_l should be near $2^{\frac{1}{3}}$ (Warner and Wilson 1976). While this model has not been extended to other R values, preliminary results on trees in a plane suggest that data should be compared with

$$R_l = R^{\frac{1}{3}} \quad (13.4$$

In Fig. 13.2, $\ln R_d$ is displayed against $\ln R$, for living trees and fossil plants, with the lines corresponding to equations (13.1) and (13.3) superimposed. The Rhyniophyte data are excluded because they are a very special case with almost perfect branching symmetry, and because almost all the cases given by Niklas

Table 13.2 Diameter of arterial branches at a bifurcation fitted to the expression $d_0^\gamma = d_1^\gamma + d_2^\gamma$. In each column the first figure refers to arteries of diameter over 200 μm, the second to arteries with smaller diameter

Organ	Number of samples		γ	
Kidney	70	73	2.75	2.53
Kidney	105	86	2.74	2.86
Kidney	140	83	2.56	2.74
Kidney	105	105	2.74	2.70
Intestine	193	140	2.63	2.68
Intestine	105	105	2.57	2.68
Intestine	105	105	2.86	2.87
Muscle	159	105	2.69	2.83
Muscle	105	105	2.86	2.87
Cerebral cortex	106	106	2.67	2.79
Basal ganglion	61	147	2.64	2.61
Pancreas	55	66	2.54	2.73
Heart	116	108	2.51	2.82
Lung	108	121	2.66	2.47

Reproduced from 'Estimation of intravascular blood pressure gradient by mathematical analysis of arterial casts', *Tohoku J. Exp. Med.* **79**, 168–198, by N. Suwa *et al.*, by permission of the Tohoku University Medical Press.

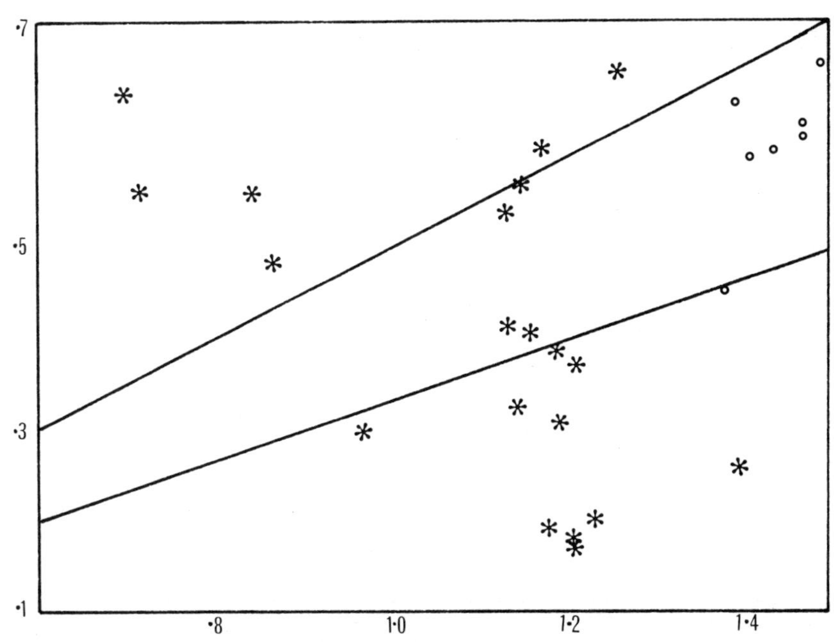

Fig. 13.2 ln R_d against ln R for tree (circles) and fossil plants (stars). The lines corresponding to $R_d = R^{\frac{1}{2}}$ and $R_d = R^{\frac{1}{3}}$ are superimposed

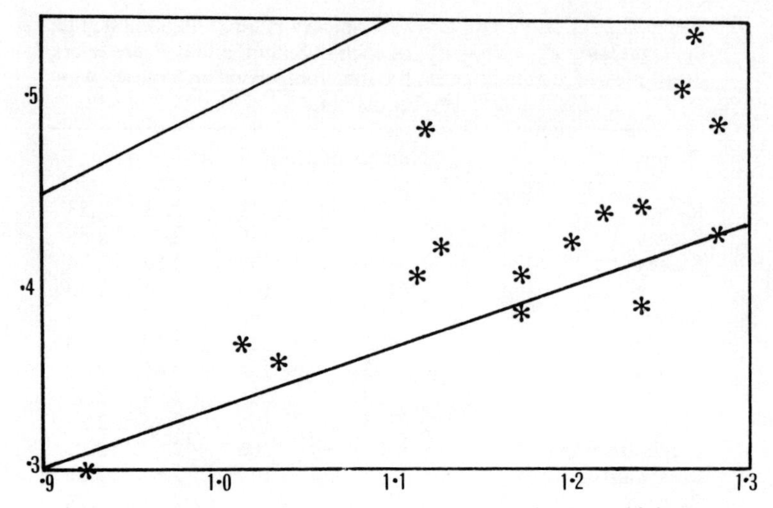

Fig. 13.3 As for Fig. 13.2, but the data are for bronchial trees

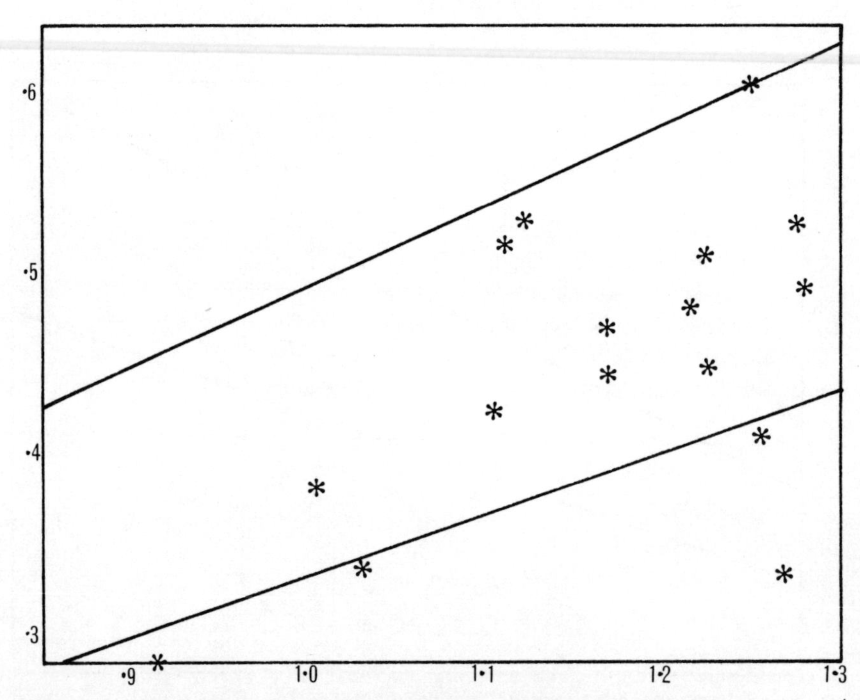

Fig. 13.4 ln R_l against ln R for bronchial trees. The lines corresponding to $R_l = R^{\frac{1}{2}}$ and $R_l = R^{\frac{1}{3}}$ are superimposed

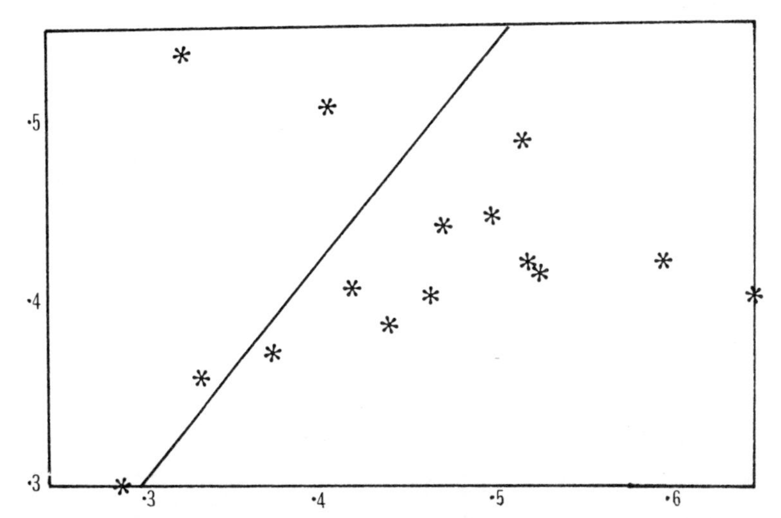

Fig. 13.5 ln R_d against ln R_l for bronchial trees. The line corresponding to $R_d = R_l$ is superimposed

(1977) have very few orders. Fig. 13.3 displays ln R_d against ln R for bronchi, with the same two lines superimposed. Fig. 13.4 displays ln R_l for bronchi, with the lines corresponding to equation (13.4) and to

$$R_l = R^{\frac{1}{2}} \tag{13.5}$$

superimposed. Finally Fig. 13.5 shows ln R_l against ln R_d for bronchi, with $R_l = R_d$ superimposed.

Plotted in this way the fossil plant data are rather scattered, and one would be rash to draw any conclusions. However, the data for bronchi suggest that the $\frac{1}{3}$ power laws have some validity. In view of the wide range of R values in the sugar maple data of Steingraeber *et al.* (1979), it would be interesting to test equation (13.3) for trees over a wider range of R than is available in the work of Barker *et al.* (1973) and of McMahon and Kronauer (1976).

Suwa *et al.* (1963) find that in several arterial systems there is a fair degree of correlation between segment length and segment diameter, which they parametrize as

$$l \propto d^a$$

with a about 0.8 for renal arteries, about 1.0 for femoral and mesenteric arteries, and about 1.2 for cerebral arteries.

13.3 SOME DETAILS OF BRANCHING

A considerable amount of detail is lost between the original data and those summarized in Table 13.1. In the remaining tables in this chapter I present more detailed information from two of the most thorough of the studies cited. Table

Table 13.3 Data for seventeen Strahler orders of human pulmonary arteries. m stands for the Strahler order, N_m for the number of branches of that order, $\bar{d}_m$ for the mean diameter of these branches, and $\bar{l}_m$ for their mean length. Note the anomalous behaviour of $\bar{l}_m$ around $m = 14$

m	N_m	$\bar{d}_m$ (mm)	$\bar{l}_m$ (mm)
1	7.3×10^7	0.013	0.13
2	2.03×10^7	0.021	0.20
3	5.64×10^6	0.034	0.25
4	1.79×10^6	0.054	0.44
5	5.67×10^5	0.086	0.65
6	1.80×10^5	0.138	0.91
7	5.81×10^4	0.224	1.38
8	1.88×10^4	0.351	2.10
9	6062	0.525	3.16
10	2290	0.850	4.69
11	675	1.33	6.60
12	203	2.09	10.5
13	66	3.65	17.9
14	20	5.82	20.7
15	8	8.06	10.9
16	3	14.8	32.0
17	1	30.0	90.5

Reproduced from 'The morphometry of the small pulmonary arteries in man', *Circ. Res.* **42**, 593–597, by K. Horsfield, by permission of the American Heart Association Inc. and the author.

Table 13.4 Data on dendrites of seventeen Purkinje cells

Numbers of segments in successive orders for cells with maximum Strahler order 7

525	171	53	18	4	2	1
421	138	45	13	4	2	1
489	162	55	17	4	2	1
503	159	48	18	5	2	1
416	141	48	12	4	2	1
443	136	40	12	4	2	1
549	174	60	19	5	2	1

Numbers of segments in successive orders for cells with maximum Strahler order 6

544	168	50	14	3	1
435	143	38	16	4	1
264	86	27	7	2	1
331	103	34	12	3	1
446	149	50	15	5	1
321	98	28	9	2	1
358	113	40	11	3	1
388	124	37	15	2	1
263	87	25	5	2	1
402	138	44	12	4	1

The rather similar mean R values quoted in Table 13.12 conceal the marked difference in branching ratios at high orders in these two sets of cells.

Reproduced from 'Network analysis of dendritic fields of pyramidal cells in the neocortex and Purkinje cells in the cerebellum of the rat', *Phil. Trans. Roy. Soc.* **B270**, 227–262, by T. Hollingworth and M. Berry, by permission of the Royal Society and of Prof. Berry.

13.3 shows the relevant quantities for the seventeen Strahler orders of the branching structure of the pulmonary arteries of a human lung, as presented by Horsfield (1978). Table 13.4 shows the number of segments belonging to branches of Strahler order up to 7 in the sixteen Purkinje cells studied by Hollingworth and Berry (1975). Unfortunately the most detailed published description of the bronchial tree is in the early papers by Horsfield and his colleagues, in which the Strahler scheme was not used.

Some remarks should be made about likely sources of error in these data. Identification of branches of Strahler order 1 is not a trivial matter. Artery and airway geometry cannot, in general, be studied *in vivo*. Exceptions include arteries in the retina (Zamir *et al.* 1979) and in the wings of bats (Wiedeman 1963). As Fig. 13.6, taken from the paper by Zamir *et al.* (1979), makes clear, even the *in vivo* examples show loss of fine branches. With plastic or resin casts of bronchial trees

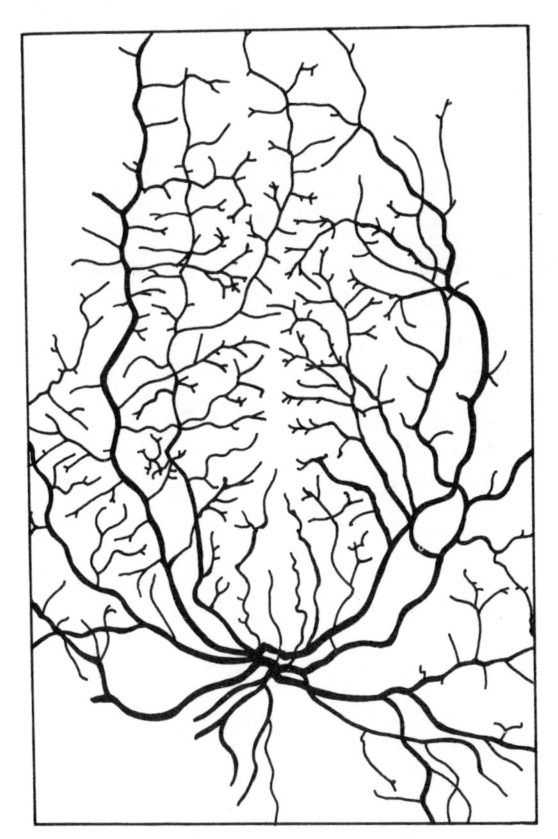

Fig. 13.6 Arteries in the human retina. Redrawn by M. F. MacDonald from an illustration in 'Arterial bifurcations in the human retina', *J. gen. Physiol.* **74**, 537–548 by M. Zamir *et al.*, by permission of the Rockefeller University Press and Dr. Zamir. © 1979 by the Rockefeller University Press

and arteries, this is even more likely. Branches can be assigned to Strahler order 1 by an arbitrary cut-off in diameter. The critical diameter is chosen so that the tree is essentially complete as far as branches of that diameter. Horsfield emphasizes that this is a reasonable convention because the mean diameter of branches in different orders correlates well with the order, in a centripetal ordering scheme.

The process of making casts by injection of resin or plastic under pressure distends the vessels. For arteries Suwa *et al.* (1963) estimate this to be equivalent to a correction

$$d_{\text{cast}} = 1.37\, d_{\text{live}}^{0.96}$$

with d_{live} measured in microns. This is proportionally a bigger effect the smaller the artery.

In living trees branches of order 1 can be unambiguously identified and mean branch diameters measured accurately. However, fossil trees have been subjected to intense pressure, causing loss of fine branches and distortion of cross-sections.

References

Barker, S. B., Cumming, G., and Horsfield, K. (1973). Quantitative morphology of the branching structure of trees, *J. theor. Biol.* **40**, 33–43.

Bowes, C., Cumming, G., Horsfield, K., Loughhead, J., and Preston S. (1982). Gas mixing in a model of the pulmonary acinus with asymmetrical alveolar ducts, *J. appl. Physiol.* **52**, 624–633.

Braverman, M. H., and Schrandt, R. G. (1967). Colony development of a polymorphic hydroid as a problem in pattern formation, *Gen. Systems* **12**, 39–51.

Cheetham, A. H., Hayek, L. A. C., and Thomson, E. (1980). Branching structure in arborescent animals: models of relative growth, *J. theor. Biol.* **85**, 335–369.

Gull, K. (1975). Mycelium branch patterns of *Thamnidium elegans, Trans. Brit. Mycol. Soc.* **64**, 321–324.

Hallé, F., Oldemann, R. A. A., and Tomlinson, P. B. (1978). *Tropical Trees and Forests: an Architectural Analysis*, Springer, Berlin.

Ho, H. H. (1978). Hyphal branching patterns in Phytophthora and other Phycomycetes, *Mycopath.* **64**, 83–86.

Holland, P. G. (1969). The maintenance of structure and shape in three mallee eucalypts, *New Phytol.* **68**, 411–421.

Hollingworth, T., and Berry, M. (1975). Network analysis of dendritic fields of pyramidal cells in the neocortex and Purkinje cells in the cerebellum of the rat, *Phil. Trans. Roy. Soc.* **B270**, 227–262.

Horsfield, K., and Cumming, G. (1968a). Morphology of the bronchial tree in man, *J. appl. Physiol.* **24**, 373–383.

Horsfield, K., and Cumming, G. (1968b). Functional consequence of airways physiology, *J. appl. Physiol.* **24**, 384–390.

Horsfield, K., Dart, G., Olson, D. E., Filley, G., and Cumming, G. (1971). Models of the human bronchial tree, *J. appl. Physiol.* **31**, 207–217.

Horsfield, K., and Cumming, G. (1976). Morphology of the bronchial tree in the dog, *Resp. Physiol.* **26**, 173–181.

Horsfield, K., Relea, F. G., and Cumming, G. (1976). Diameter, length and branching ratios in the bronchial tree, *Resp. Physiol.* **26**, 351–356.

Horsfield, K. (1977). Postnatal growth of the dog's bronchial tree, *Resp. Physiol.* **29**, 185–191.

Horsfield, K. (1978). The morphometry of the small pulmonary arteries in man, *Circ. Res.* **42**, 593–597.

Horsfield, K., and Thurlbeck, A. (1981). Relation between diameter and flow in branches of the bronchial tree, *Bull. math. Biol.* **43**, 681–691.
Horsfield, K., and Gordon, W. I. (1981). Morphometry of pulmonary veins in man, *Lung* **159**, 211–218.
Horsfield, K., Kemp, W., and Phillips, S. (1982). An asymmetrical model of the airways of the dog lung, *J. appl. Physiol.* **52**, 21–26.
McMahon, T. A., and Kronauer, R. F. (1976). Tree structures—deducing properties of mechanical design, *J. theor. Biol.* **59**, 443–466.
Murray, C. D. (1927), A relationship between circumference and weight and its bearing on branching angles, *J. gen. Physiol.* **10**, 725–729.
Niklas, K. J. (1978). Morphometric relationships and rate of evolution among Palaeozoic vascular plants, in *Evolutionary Biology* (edited by M. K. Hecht), Plenum, New York, pp. 509–543.
Oohata, S., and Shidei, T. (1971). Studies in the branching structures of trees. I Bifurcation ratio of trees in Horton's law, *Jap. J. Ecol.* **21**, 7–14.
Parker, H., Horsfield, K., and Cumming, G. (1971). Morphology of distal airways in the human lung, *J. appl. Physiol.* **31**, 386–391.
Phalen, R. F., Yeh, H. C., Schum, G. M., and Raabe, O. G. (1978). Application of an idealised model to morphometry of the mammalian tracheobronchial tree, *Anat. Rec.* **190**, 167–176.
Raabe, O. G., Yeh, H. C., Schum, G. M., and Phalen, R. F. (1976). Tracheobronchial geometry: human, dog, rat, hamster, Lovelace Foundation, Albuquerque.
Shinozaki, K., Yoda, K., Hozami, K., and Kira, T. (1964a). A quantitative analysis of plant form: the pipe model theory. I Basic analysis, *Jap. J. Ecol.* **14**, 97–105.
Shinozaki, K., Yoda, K., Hozami, K., and Kira, T. (1964b). A quantitative analysis of plant form: the pipe model theory. II Further evidence of the model and applications to forest ecology, *Jap. J. Ecol.* **14**, 133–139.
Singhal, S., Henderson, R., Horsfield, K., Harding, K., and Cumming, G. (1973). Morphometry of the human pulmonary arterial tree, *Circ. Res.* **33**, 190–197.
Steingraeber, D. A., Kascht, L. J., and Frank, D. H. (1979). Variation of shoot morphology and bifurcation ratio in sugar maple (*Acer saccharum*) saplings, *Am. J. Bot.* **66**, 441–445.
Suwa, K., Niwa, T., Fukusawa, H., and Sasaki, Y. (1963). Estimation of intravascular blood pressure by mathematical analysis of arterial casts, *Tohoku J. exp. Med.* **79**, 168–198.
Thurlbeck, A., and Horsfield, K. (1980). Branching angles in the bronchial tree related to the order of the branch, *Resp. Physiol.* **41**, 173–181.
Tomlinson, P. B. (1978). Some qualitative and quantitative aspects of New Zealand divaricating shrubs, *N.Z.J. Bot.* **16**, 299–309.
Warner, W. H., and Wilson, T. A. (1976). Distribution of the end points of a branching network with decaying branch length, *Bull. math. Biol.* **38**, 219–237.
Whitney, G. C. (1976). The bifurcation ratio as an indicator of adaptive strategy in woody plant species, *J. Torrey Bot. Club* **103**, 67–72.
Wiedeman, M. P. (1963). Dimensions of blood vessels from distributing artery to collecting vein, *Circ. Res.* **12**, 375–378.
Williamson, M. E. (1967). The venous and biliary systems in the bovine liver, Master's thesis, Cornell University.
Woldehirst, Z., and Horsfield, K. (1978). Diameter, length and branching ratios of the upper airways in the dog lung, *Resp. Physiol.* **33**, 213–218.
Woldenberg, M. J. (1972). The average hexagon in spatial hierarchies, in *Spatial Analysis in Geomorphology* (edited by R. J. Chorley), Methuen, London, pp. 323–352.
Zamir, M., Medeiros, J. A., and Cunningham, T. K. (1979). Arterial bifurcation in the human retina, *J. gen. Physiol.* **74**, 537–548.

Chapter 14

The lung as a space-filling tree

The bronchial tree and the pulmonary arterial tree serve to supply air and blood to the alveoli, where gas transfer takes place. There is an advantage in a uniform distribution of the alveoli throughout the lung volume. This has certain implications for the geometry of these two trees, which can be regarded as space-filling structures. To understand this it is necessary first to be clear in what sense a tree, with branches of negligible volume compared with the region of space in which they ramify, can be described as filling that region.

14.1 SYMMETRICAL BRANCHING AT 90°

It is usual to think of the dimension of a space in terms of the number of orthogonal axes that are needed to specify the direction of an arbitrary straight line in the space. However, other definitions of dimension are possible. One in particular allows us to define the dimension of a branching structure of lines, as distinct from the dimension of a line. An informal definition of similarity dimension can be given following Chapter 6 of Mandelbrot (1982). A line segment can be divided up into N identical segments, each scaled down by a factor

$$r_1(N) = N^{-1}$$

A rectangle can be divided into N identical rectangles, each scaled down by a factor

$$r_2(N) = N^{-\frac{1}{2}}$$

A cube can be divided into N identical cubes, each scaled down by a factor

$$r_3(N) = N^{-\frac{1}{3}}$$

The similarity dimension, which in these cases is identical with the usual dimension, is defined as

$$D = -\ln N/\ln r_D(N) \qquad (14.1)$$

A binary tree can be assigned a similarity dimension if it possesses a property of

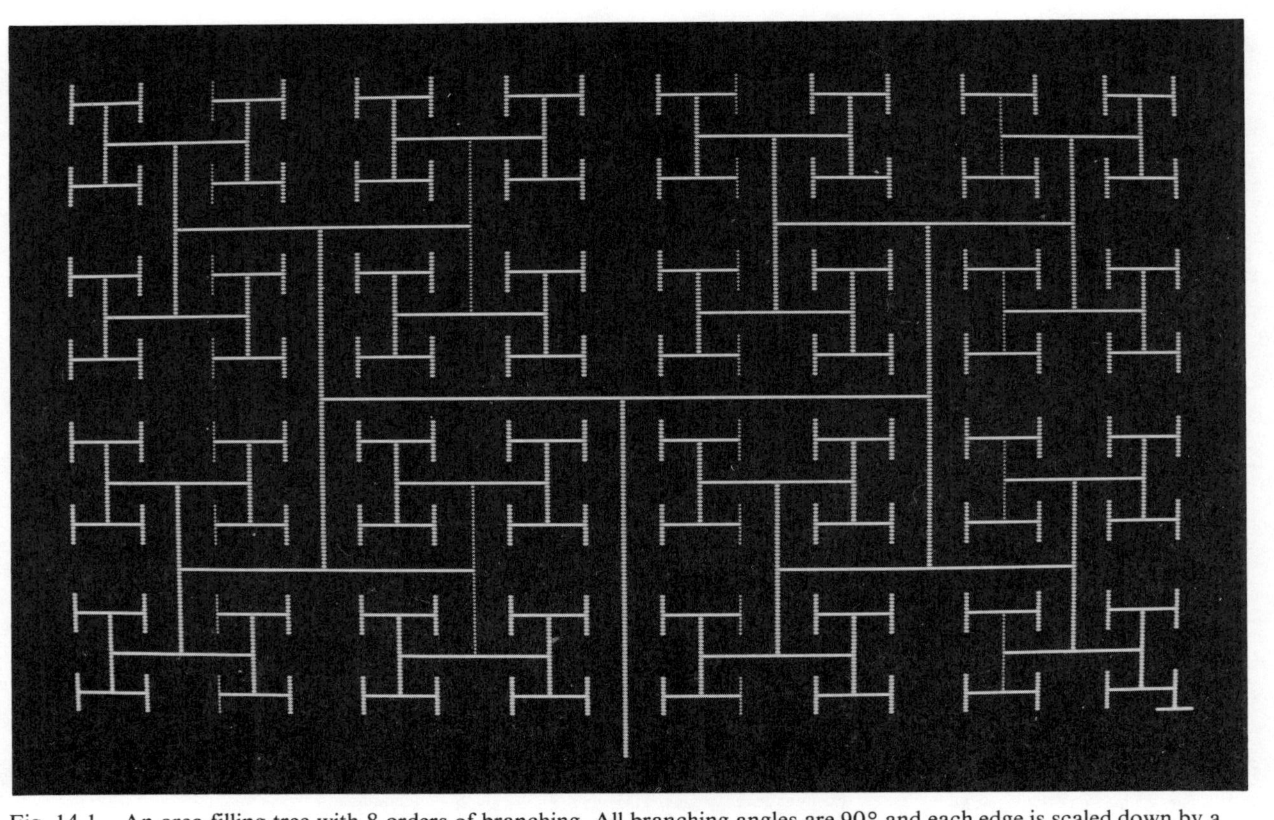

Fig. 14.1 An area-filling tree with 8 orders of branching. All branching angles are 90° and each edge is scaled down by a factor $2^{-\frac{1}{2}}$ from its parent edge. Photographed from the screen of an Apple II computer

self-similarity. If $D = 2$ the tree fills a two-dimensional region, in this sense: given an arbitrary point P in the region, we can place an end point of the tree as near as we like to P by taking a sufficient number of bifurcations. In the same way, if $D = 3$, the tree can fill a three-dimensional volume.

Fig. 14.1, taken from the screen of an Apple II computer, shows a symmetrical binary tree with all branching angles equal to 90°, and with the ratio $R_l = 2^{\frac{1}{2}}$, so that successive edges are reduced in length by a factor $2^{-\frac{1}{2}}$. The first eight orders of branching are shown. The first bifurcation is at the centre of a certain rectangular region, the edges of which lie just beyond the limits of the pattern shown. After two steps there is a bifurcation at the centres of each of the four rectangles into which the large rectangle can be symmetrically divided. This process continues, and the tree in consequence has the same similarity dimension as the rectangle. Fig. 14.2 shows the first four bifurcations of a symmetrical binary tree which fills a cube. All branching angles are 90°. The successive planes, containing a parent and both daughter edges, lie parallel to the XY, YZ, ZX planes in turn. The ratio $R_1 = 2^{\frac{1}{3}}$, so that the successive edges are reduced in length by a factor $2^{-\frac{1}{3}}$. This tree has similarity dimension 3. Cohn (1954) pointed out that this could be used as a very crude model of the whole arterial tree of a mammal. In a subsequent paper Cohn (1955) used this kind of model for compact

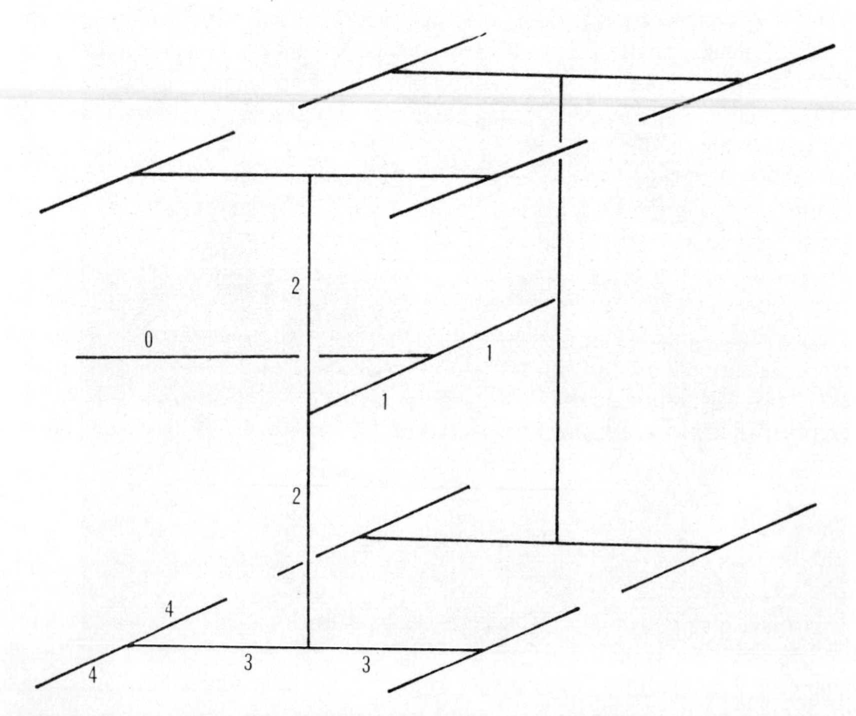

Fig. 14.2 A volume-filling tree with four orders of branching. The scaling factor is $2^{-\frac{1}{3}}$. The branching angles are 90°. Edges are in turn parallel to the x, y, and z axes

parts of the arterial tree, combining it with an asymmetrical model of the aorta and the main arteries.

In his first paper Cohn used the data summarized by Green (1950), which related to the arterial system of a 13 kg dog. Cohn estimated the equivalent cube to have sides 0.23 m. Estimating the volume of tissue that can be supplied by one capillary as 2.7×10^{-12} m^3, he concluded that this model requires 33 bifurcations in a path from aorta to capillary, corresponding to eleven subdivisions of the cube symmetrically into cubes of one-eighth the volume. He noted that Green (1950) lists nine major types of vessel from the aorta to the capillaries. Later Iberall (1967) listed about eleven stages of branching, as discussed in Section 11.3, but in a very different way from Cohn. Whereas Cohn requires three successive apical branchings in each of his eleven subdivisions of the cube, Iberall requires at each stage a branch with several lateral branches as well as a final apical branching.

14.2 SYMMETRICAL BRANCHING AT ANY ANGLE

Now it is clear that it would be more reasonable to apply this style of model in a single compact organ, such as the lung, than in the whole body of a mammal. But the bronchial tree, and the pulmonary arterial tree, differ in several ways from the idealized space-filling structure described above. They are rather asymmetrical; when Strahler ordering is adopted they obey Horton's law with $R = 3$ rather than $R = 2$. Branching angles are not usually equal at a bifurcation, and tend to be considerably smaller than 90°. The bronchi and arteries have finite volumes and thus the tree itself makes some of the lung volume unavailable for alveoli. An investigation of the space-filling characteristics of these trees is thus difficult. A step towards such an investigation was made in the work of Warner and Wilson (1976) in terms of a random walk with shrinking steps. The branches are treated as of zero radius; an excluded volume random walk would be a much harder problem. The model is symmetrical, with $R = 2$ and with equal branching angles $\pm\alpha$ at every bifurcation, and with a set value of R_l, so that successive branches shrink by a constant factor. This leaves only one property to be assigned in a random manner, the orientation of the plane of bifurcation. Restrictions on the randomness of this orientation, in the first few bifurcations of the lung airways, have been described by Chen *et al.* (1980). von Hayek (1960) notes that the angle between the two daughter branches at a bifurcation, in human lung airways, is more acute the larger the branches. Thurlbeck and Horsfield (1980) present detailed data on a similar feature of the lung airways of a dog. They find that over six successive Strahler orders this angle (corresponding to 2α in the model discussed here) falls from 80° to 60°.

Methods are available, as for example described by Kac (1959), for three-dimensional random walk with steps turning aside by $\pm\alpha$, and with $R_l = 1$. In this symmetrical model, for $\alpha = 90°$, $R_l = 2^{\frac{1}{3}}$ gives a tree that fills a cube. Any larger R_l causes branches to shrink too rapidly, leaving gaps. Any smaller α, for $R_l = 2^{\frac{1}{3}}$, also leaves gaps. The interest is therefore in the range $1 < R_l < 2^{\frac{1}{3}}$. Warner and Wilson adapt the method described by Kac (1959) to allow $R_l \neq 1$.

A coordinate system $\{X_1, X_2, X_3\}$ is chosen with the axis OX_1 along the direction of the main stem, the origin O lying at the beginning of that stem. Taking l_0 as the length of the main stem, the first bifurcation is at $\{l_0, 0, 0\}$. After m bifurcations an end point has coordinates

$$\boldsymbol{R}^m = \{X_1^m, X_2^m, X_3^m\} \tag{14.2}$$

There are 2^m such endpoints. The aim is to calculate, for $m \to \infty$, the mean values and variances of X_r^m. (This is called an asymptotic calculation because it goes to this limit, $m \to \infty$.)

Successive vectors are linked by

$$\begin{aligned} \boldsymbol{R}^m &= \boldsymbol{R}^{m-1} + \boldsymbol{r}^m \\ &= l_0 \sum_{p=0}^{m} \gamma^p \boldsymbol{e}^p \end{aligned} \tag{14.3}$$

Here $\gamma = R_l^{-1}$, and $\boldsymbol{e}^p$ is the unit vector along the pth branch. To specify a rule for the orientation of successive branches it is necessary to define a set of orthogonal unit vectors $\{\boldsymbol{e}^p, \boldsymbol{f}^p, \boldsymbol{g}^p\}$, where $\boldsymbol{f}^p$ is in the bifurcation plane of the pth branch and its parent branch. The first of these sets, $\{\boldsymbol{e}^0, \boldsymbol{f}^0, \boldsymbol{g}^0\}$ lie along the OX_1, OX_2 and OX_3 directions respectively.

The vectors $\{\boldsymbol{e}^p, \boldsymbol{f}^p, \boldsymbol{g}^p\}$ are obtained from the vector $\{\boldsymbol{e}^0, \boldsymbol{f}^0, \boldsymbol{g}^0\}$ by a sequence of rotations. At each stage one first rotates about $\boldsymbol{e}^{p-1}$, starting from the previous bifurcation plane $\boldsymbol{e}^{p-1}, \boldsymbol{f}^{p-1}$, to give a new bifurcation plane; this defines a rotation angle ϕ^p. Then one rotates by an angle θ^p about the normal $\boldsymbol{g}^p$ to the new bifurcation plane. Two different probabilistic treatments give somewhat similar results. In the first all the ϕ^p are set equal to a value ψ, where $-\pi/2 \leqslant \psi \leqslant \pi/2$. In the second the ϕ^p are distributed uniformly in the range $-\pi/2$ to $\pi/2$. In each the angles θ^p are set equal to $\pm\alpha$ with equal probability. Formally

$$\begin{aligned} \begin{bmatrix} \boldsymbol{e}^p \\ \boldsymbol{f}^p \\ \boldsymbol{g}^p \end{bmatrix} &= \begin{bmatrix} \cos\theta^p & \sin\theta^p & 0 \\ -\sin\theta^p & \cos\theta^p & 0 \\ 0 & 0 & 1 \end{bmatrix} \begin{bmatrix} 1 & 0 & 0 \\ 0 & \cos\phi^p & \sin\phi^p \\ 0 & -\sin\phi^p & \cos\phi^p \end{bmatrix} \begin{bmatrix} \boldsymbol{e}^{p-1} \\ \boldsymbol{f}^{p-1} \\ \boldsymbol{g}^{p-1} \end{bmatrix} \\ &= A(\theta^p, \phi^p) \begin{bmatrix} \boldsymbol{e}^{p-1} \\ \boldsymbol{f}^{p-1} \\ \boldsymbol{g}^{p-1} \end{bmatrix} \end{aligned} \tag{14.4}$$

and

$$\begin{bmatrix} \boldsymbol{e}^m \\ \boldsymbol{f}^m \\ \boldsymbol{g}^m \end{bmatrix} = \prod_{p=1}^{m} A(\theta^p, \phi^p) \begin{bmatrix} \boldsymbol{e}^0 \\ \boldsymbol{f}^0 \\ \boldsymbol{g}^0 \end{bmatrix} \tag{14.5}$$

The coordinates of the endpoints of the mth generation are elements of the first row of the matrix

$$l_0\left[I + \gamma A(\theta^1, \phi^1) + \cdots + \gamma^m \prod_{p=1}^{m} A(\theta^p, \phi^p)\right] \tag{14.6}$$

where I is the unit 3×3 matrix.

The probabilistic nature of the calculation leads to two consequences. The first is that, whereas the lung resembles the ideal cube-filling tree in filling up a clearly defined volume, the model creates a family of random walks, occupying a certain average volume defined as the product of the variances of the endpoint coordinates. The second is that the concept of uniformly filling a volume with end-points has to be examined. Warner and Wilson (1976) point out that a space cannot be filled uniformly by the endpoints of a set of random trees if the sum of the volumes (in the diffuse sense just specified) occupied by trees starting at any of the pth bifurcation points is less than the sum of the volumes occupied by the complete trees. They have checked that uniform occupancy is ensured in the cases of interest.

For each of a set of values of γ from 1 to $2^{-\frac{1}{3}}$ these authors present a curve of volume against branching angle α at fixed l_0. This is reproduced in Fig. 14.3.

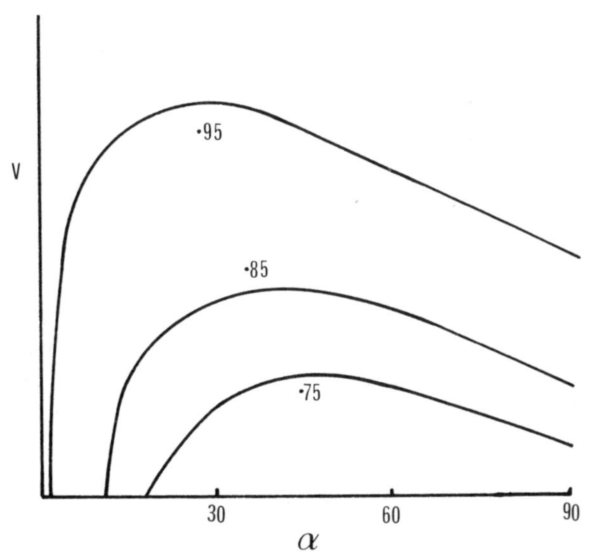

Fig. 14.3 The volume V occupied by a symmetrical tree plotted against branching angles α, with scaling factor $\gamma = 0.75$, 0.85, and 0.95. Reproduced from 'Distribution of end points of a branching network with decaying branch length', *Bull. Math. Biol.* **38**, 219–237, by W. H. Warner and T. A. Wilson, by permission of Pergamon Press. © 1976 by Pergamon Press

When the lung is required to fill a certain volume V, at given γ, then the branching angle giving the maximum on this curve allows this for the least value of l_0

$$l_0 = L_0(\alpha_{max}, \gamma)$$

To choose between all the pairs α_{max}, γ it is not enough to pick the lowest L_0. The actual volume of the branches, which must be as small as possible to leave as much space as possible for the alveoli, is

$$V^B = \frac{L_0(\alpha_{max}, \gamma)}{4} \sum_{p=0}^{M} \gamma^p 2^p \bar{d}_p^2 \tag{14.7}$$

where $\bar{d}_p$ is the mean diameter in the pth order, and M is the total number of generations of branching (Weibel orders, as defined in Section 12.1). From Fig. 14.3, it is apparent that raising γ lowers L_0, but from equation (14.7) it is apparent that raising γ raises the other factor in V^B. A minimum V^B is reached for

$$\gamma = 0.85, \ \alpha_{max} = 35°$$

The lung volume can be filled, in the diffuse sense appropriate to this model and with economy in terms of V^B, by a symmetrical binary tree with fixed γ and fixed α, and random orientation of bifurcation planes, by selecting these values of γ and α. For comparison, $2^{-\frac{1}{3}}$ is 0.794, the value appropriate for $\alpha = 90°$.

14.3 PRELIMINARY IDEAS ON ASYMMETRY

It would be of considerable interest to extend this style of analysis to $R = 3$ trees. It is tempting to conjecture that values of γ near $R^{-\frac{1}{3}}$, rather than near $2^{-\frac{1}{3}}$, are appropriate for space filling. There are signs, in some 2-dimensional simulations to be described, that this is plausible. This is why I have drawn the line $R_l = R^{\frac{1}{3}}$ in Fig. 13.4. It is also possible that space filling characteristics may be sensitive to R, and a range from $R = 2$ to $R = 4$ might profitably be investigated. While Warner and Wilson (1976) can get quite far analytically in the case $R = 2$, extensions to other R values are liable to involve heavy computation. As a preliminary to an investigation of this kind, I have simulated a number of two-dimensional trees on an Apple computer. These are of two types. One is suggested by the use of Strahler ordering and the fact that empirically R is about three. The other uses a model suggested by the Horsfield ordering scheme. To get the simplest possible tree with $R = 3$, I take as basic unit a Y with a side arm. The slanting branches of the Y are at angles $\pm\alpha$ with the stem, and the side arm is at right angles to the stem, with random orientation to left or right. but with the branching point always half-way along the stem. All three branches are reduced in length by the same factor, compared with the length of the stem. An example, with six orders of branching, is shown in Fig. 14.4. It is clear that there are two features that should both be avoided if possible, gaps and tangling. To get the simplest case with $R = 4$, with no random element, I take a Y with a symmetrical cross arm, that is a Y on top of a T. There are indications, particularly in the case of $R = 4$, that a

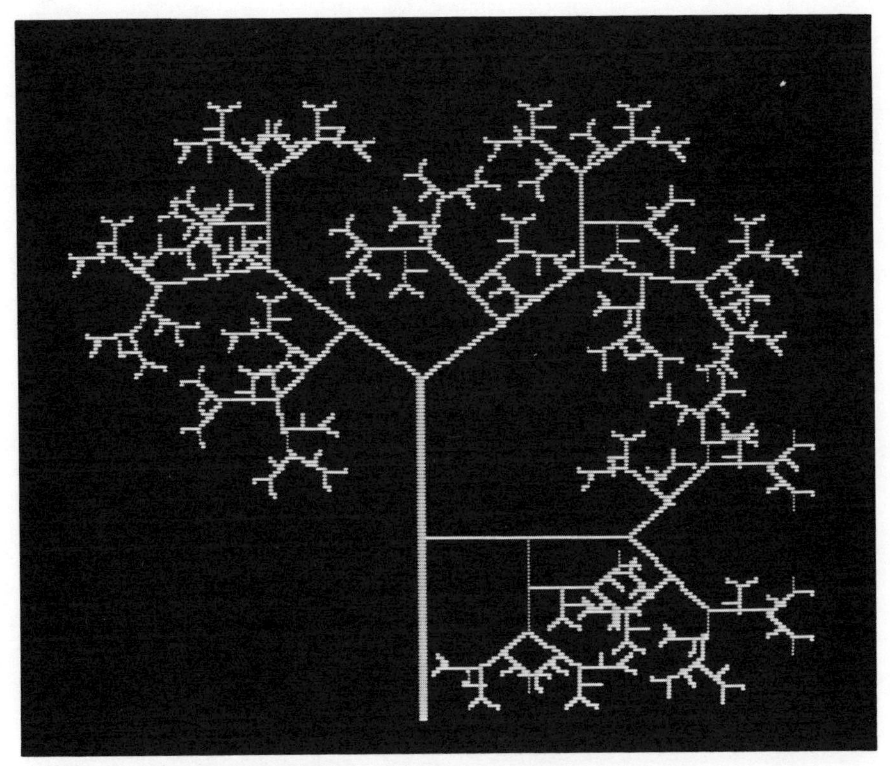

Fig. 14.4 An asymmetrical tree with $R = 3$. Photographed from the screen of an Apple II computer

scaling factor $R^{-\frac{1}{2}}$ rather than $2^{-\frac{1}{2}}$, is suitable if one is to avoid gaps and tangling. Extrapolation to three dimensions is, however, to be done with caution since tangling, which results when $2^{-\frac{1}{2}}$ is used, is likely to be less of a problem in three dimensions.

An equally simple kind of tree, with a somewhat more 'organic' look to it, is illustrated in Fig. 14.5. The basic unit is an asymmetrical Y, with the branch which comes off at $\pm 2\alpha$ scaled down more strongly than that which comes off at $\mp\alpha$. It is possible to relate the scaling factor in a three-dimensional model of this kind to the empirical values for bronchial trees, when the Horsfield ordering method is used. This method is centripetal like Strahler ordering, but assigns order $m + 1$ to the parent edge which gives rise to edges of order m and $n < m$ (see Fig. 12.6).

Horsfield *et al.* (1971, 1982) describe how to approximate a tree ordered in this way by a model in which all bifurcations, down to those giving rise to branches of order $\Delta + 1$, are of the type

$$m + 1 \rightarrow m, m - \Delta \tag{14.8}$$

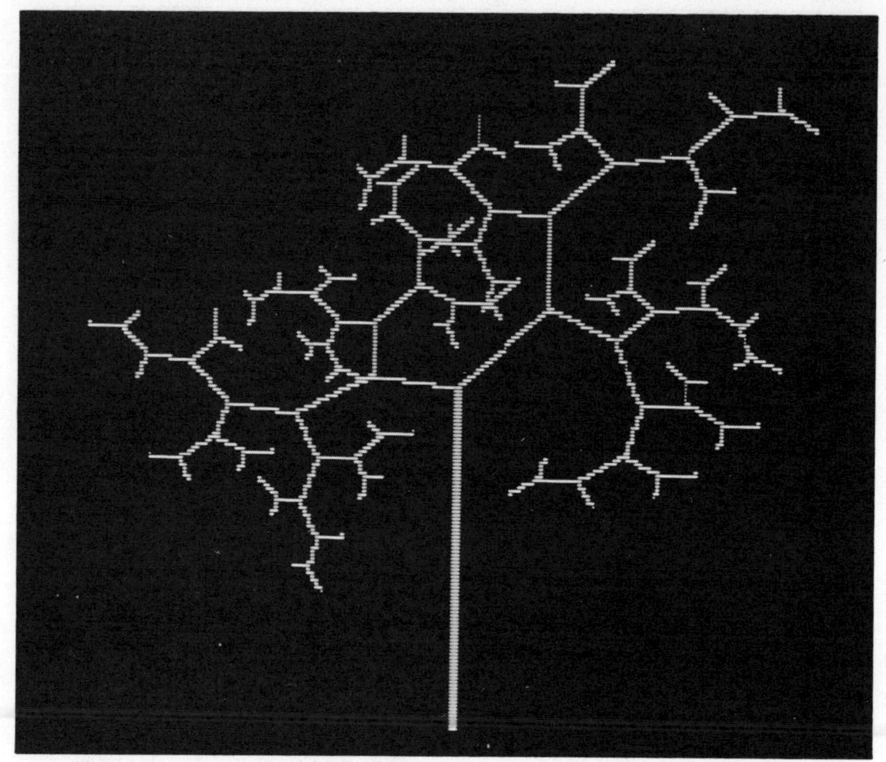

Fig. 14.5 An asymmetrical tree of the type suggested by the regular model of Horsfield. Photographed from the screen of an Apple II computer

This is a generalization, suitable for asymmetrical trees, of the Weibel symmetrical model, in which $\Delta = 0$. Using Horsfield ordering the empirical branching ratio for the human lung is about 1.4 (Horsfield *et al.* 1971) which can be realized in the model with $\Delta = 3$. If one adopts a scaling factor γ in length from one Horsfield order to the next, the asymmetrical Y unit has one arm shorter than its stem by γ, and the other by γ^4. My two-dimensional model combines this with the reasonable assumption that the shorter branch comes off at the larger angle.

Three-dimensional regular models of either of these types, with the bifurcation planes oriented at random, but otherwise having the random features of the branching removed, could be a useful intermediary between the excessively symmetric model of Warner and Wilson (1976) and a realistic random asymmetrical tree. The Horsfield one probably has the greater potential for analytical treatment. (The regularity of equation (14.8) cannot be carried out to all orders, as illustrated by Fig. 14.6. But for an asymptotic treatment like that of Warner and Wilson this is not a problem.)

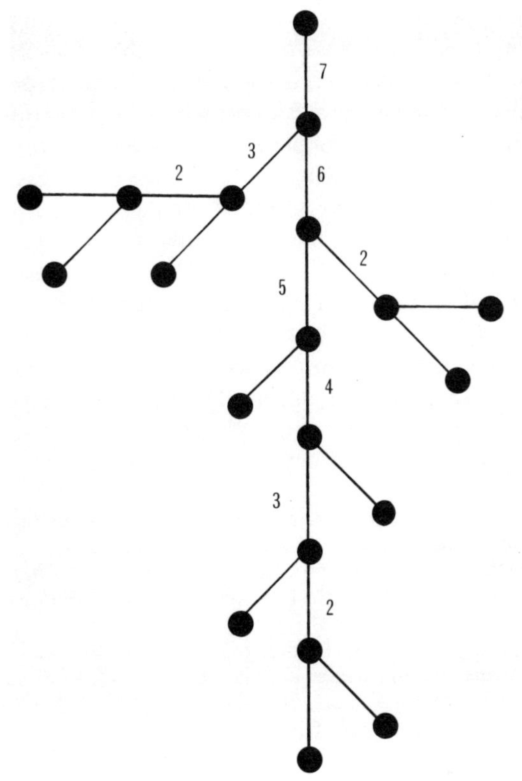

Fig. 14.6 The truncation of a regular model of the Horsfield type

14.4 FILLING A SURFACE WITH LEAVES

Turning to botanical trees, there is no advantage in filling a volume with leaves if those near the centre receive little sunlight. Rather the leaves should fill a shell between two surfaces. One could idealize to a branching structure which uniformly fills a surface with endpoints. Arguments of this kind have been given by Woldenberg (1972), Honda and Fisher (1978), and Fisher and Honda (1979). Woldenberg rather overstresses the role of surface filling, claiming it as the basis for the tree structure of stream networks—where branches fill a surface in which they lie, botanical trees—where branches are transverse to the surface filled by their endpoints, and arterial and bronchial trees—where volume filling is a more natural approach, except possibly in some special cases such as the kidney. Woldenberg (1972) does not discuss the question of a power law relationship between R and R_l, preferring to express empirical results in terms of the ratio of these two members.

Fisher and Honda (1979) discuss the species *Terminalia catappa* in which there are several roughly horizontal tiers of branches, each having about five main branches emerging from the vertical trunk (in fact that mean number is 4.73). In each tier the subsequent branching is by bifurcation. Since only a few Strahler orders are involved, the asymptotic approach of Warner and Wilson is not appropriate. Fisher and Honda simulate a tier by m main branches of Strahler order 3, with given R_l and branching angles. At the tips of each order 1 branch they draw a disc, representing leaves, and where these discs overlap they replace them by Dirichlet polygons. (Around each tip the Dirichlet polygon contains all points nearer to that tip than to any other.) Adding the areas of these polygons gives the effective leaf area of the tier. Shading by higher tiers is ignored. The most efficient value of m for space filling is five. The most efficient value of R_l, however, is not in accord with observed values. Fisher and Honda stress that mechanical considerations may dominate this aspect of the geometry.

References

Chen, W. J. R., Shiah, D. S. P., and Wang, C. S. (1980). A three dimensional model of the upper tracheo-bronchial tree, *Bull. math. Biol.* **42**, 847–859.

Cohn, D. (1954). Optimal systems I. The vascular system, *Bull. math. Biophys.* **16**, 59–74.

Cohn, D. (1955). Optimal systems II. The vascular system, *Bull. math. Biophys.* **17**, 219–227.

Fisher, J. B., and Honda, H. (1979). Branch geometry and effective leaf area: a study of terminalia branching pattern, *Amer. J. Bot.* **66**, 633–644 and 645–655.

Green, H. (1950). Circulatory system: physical principles, in *Medical Physics* Vol. 2 (edited by O. Glasser), Yearbook Publishers, New York, pp. 228–251.

Honda, H., and Fisher, J. B. (1978). Tree branching angle: maximising effective leaf area, *Science* **199**, 888–889.

Horsfield, K., Dart, G., Olson, D. E., Filley, G. F., and Cumming, G. (1971). Models of the human bronchial tree, *J. appl. Physiol.* **31**, 207–217.

Horsfield, K., Kemp, W., and Phillips, S. (1982). An asymmetrical model of the airways of the dog lung, *J. appl. Physiol.* **52**, 21–26.

Iberall, A. S. (1967). Anatomy and steady flow characteristics of the arterial system, with an introduction to its pulsatile characteristics, *Math. Biosciences* **1**, 375–395.

Kac, M. (1959). *Probability and related Topics in Physical Sciences*, Interscience, New York.

Mandelbrot, B. B. (1982). *The Fractal Geometry of Nature*, Freeman, San Francisco.

Thurlbeck, A., and Horsfield, K. (1980). Branching angles in the bronchial tree related to the order of the branch, *Resp. Physiol.* **41**, 173–181.

von Hayek, H. (1960). *The Human Lung*, Hafner, New York.

Warner, W. H., and Wilson, T. A. (1976). Distribution of endpoints of a branching network with decaying branch length, *Bull. math. Biol.* **38**, 219–237.

Woldenberg, M. J. (1972). The average hexagon in spatial hierarchies, in *Spatial Analysis in Geomorphology* (edited by R. J. Chorley), Methuen, London, pp. 323–352.

Chapter 15

Branching diameter ratios and branching angles: biophysics

This chapter is concerned in part with the implications of the comment by d'Arcy Thompson quoted at the beginning of Part III. There is a conflict of design aims in the arterial system. Vessels with large diameters offer less resistance to the flow of blood, but they require the maintenance of a large volume of circulating blood. Since air is cheap it might seem that volume is less critical for the lung airways. But work has to be done moving air through the conducting airways, and much of this air does not get to the alveoli. Also there is a correlation between diameter and wall thickness in the vessels, so that the volume of wall material needed increases roughly as the volume of the vessels. Again the space filling distribution of the alveoli is hampered if too much of the lung volume is taken up by the conducting airways.

15.1 THE RATIO R_d

The common mathematical framework for the topics discussed here is the concept of a cost function $F(x_i)$ to be minimized by suitable choice of values for one or more variables x_i. If only one variable x is free to vary, elementary calculus tells us that the minimum is attained for $x = x^0$, where

$$\left.\frac{dF}{dx}\right|_{x=x^0} = 0$$

$$\left.\frac{d^2F}{dx^2}\right|_{x=x^0} > 0 \qquad (15.1)$$

When several variables are involved it may be necessary to employ the methods of the calculus of variations, as discussed in a biological context by Rosen (1967). But in the only application in this chapter in which two variables are simultaneously free to change, we shall require merely some extensions of equations (15.1) involving first and second partial derivatives (see Appendix 4).

Cost functions incorporating the balance of volume and resistance have been discussed by various authors, at least since the work of Murray (1926a). I shall follow an account given by Wilson (1967) in the context of the lung airways. Wilson expresses his analysis in terms of the symmetrical model of Weibel (1963) with $R = 2$. However, it can readily be extended to an asymmetric bronchial tree ordered by the Strahler method, with $R > 2$. Wilson expresses his results in terms of entropy production. However, as he treats the lung as operating at uniform constant temperature T, converting rate of entropy production to rate of working merely requires multiplication by T.

Horsfield (1977) uses what is ostensibly an argument concerning entropy production in discussing resistance to air flow in the bronchial tree. He relies on an analogy between the lung (in which muscles pump gas through tubes) and a river (in which gravity pulls liquid along a channel). This leads him to ignore completely the volume aspect of the problem. Also he relies on certain results of Leopold and Langbein (1962) on the profile of a river, which in turn depend on a formal analogy with thermodynamics which does not bear close inspection.

Wilson (1967) writes the total ventilation $\dot{V}$ as

$$\dot{V} = \dot{V}_A + \nu V \tag{15.2}$$

where ν is the frequency of breathing, V the volume of the airways, and the alveolar ventilation $\dot{V}_A$ is taken to be constant. In terms of the mean length l_m and mean diameter d_m of branches of Strahler order m, and of the Strahler order M of the trachea, this becomes

$$\dot{V} = \dot{V}_A + \frac{\pi\nu}{4} \sum_{m=1}^{M} d_m^2 l_m R^{M-m} \tag{15.3}$$

provided one assumes that Horton's law applies. The other significant quantity in the model is the pressure drop along a branch. It is important to note that, whereas the volume is an extensive variable, and requires a sum over all branches, the total pressure drop ΔP is an intensive variable, and only requires a sum over Strahler orders. On the assumption of laminar flow (see Appendix 3) this quantity is

$$\Delta P = \sum_{m=1}^{M} \Delta P_m = \frac{128\mu}{\pi} \sum_{m=1}^{M} \frac{Q_m l_m}{d_m^4} \tag{15.4}$$

Here μ is the viscosity of air, and Q_m is the flow through a typical branch of Strahler order m. Q_m is approximated by

$$\dot{V} R^{m-M}$$

where the ventilation $\dot{V}$ is the flow through the trachea, and it is assumed that the flow through any branch is proportional to the number of branches of order 1 which it subtends. This is roughly equivalent to letting the flows to all the alveoli be equal.

Wilson takes the rate of working of the respiratory muscles to be proportional to the oxygen consumption by these muscles in unit time, U. He seeks to minimize this quantity with respect to variation of d_m, by setting

$$\left(\frac{\partial U}{\partial \dot{V}}\right)_{\Delta P}\frac{\partial \dot{V}}{\partial d_m}+\left(\frac{\partial U}{\partial \Delta P}\right)_{\dot{V}}\frac{\partial \Delta P}{\partial d_m}=0 \tag{15.5}$$

(Here, for example, the subscript ΔP means that ΔP is held constant.). Equations (15.3) and (15.4) and the assumed form of Q_m allow the second factor in each pair in equation (15.5) to be evaluated

$$\frac{\partial \dot{V}}{\partial d_m}=\frac{\nu\pi}{2}d_m l_m R^{M-m} \tag{15.6}$$

$$\frac{\partial \Delta P}{\partial d_m}=-\frac{512\mu\dot{V}}{\pi}l_m d_m^{-5}R^{m-M} \tag{15.7}$$

Since both $\dot{V}$ and ΔP depend in the same way on l_m, no knowledge of l_m is required. It has to be assumed that R is determined in some way independent of the rate of working. Without any information about the values taken by the first factor in each of the pairs in equation (15.5), it is clear that the outcome of minimizing the rate of working must be that

$$d_m \propto R^{(M-m)/3} \tag{15.8}$$

On recalling how R_d is defined in Section 12.3, this means that

$$R_d = R^{\frac{1}{3}} \tag{15.9}$$

It is, of course, important that the other factors in equation (15.5) can be obtained, and Wilson (1967) takes them from the work of Campbell *et al.* (1957). The data are obtained in 'black box' experiments, in which a subject breathes through either an external vessel of variable volume, or an external device of variable resistance. So one can separately determine the effect on U of either changing $\dot{V}$ with ΔP held constant, or changing ΔP with $\dot{V}$ held constant.

To sum up, in Wilson's model the overall rate of working is minimized by separately minimizing the contribution from each order of branching. The volume terms increase with d_m, the pressure drop terms diminish as d_m increases, and the net effect, if Horton's law applies, is that $R_d = R^{\frac{1}{3}}$. For the case $R = 2$ the result $R_d = 2^{\frac{1}{3}}$ has been derived by Horsfield and Cumming (1967) in a manner which may seem at first sight contradictory to that of Wilson (1967). They minimize the sum of a volume term and a resistance term, resistance being proportional to d_m^{-4} without the flow factor Q_m. However, when many branches of a given order m are treated as parallel paths for flow, the reciprocals of the resistances are additive. So the net resistance in order m is reduced by the same factor R^{m-M} which appears in Wilson's treatment of ΔP.

The assumption of laminar flow can be criticized both for air and for blood. However, so long as the pressure drop can be expressed in the form

$$Q_m l_m f(d_m)$$

the argument can be carried through precisely as above, with a conclusion that differs only in detail. For example, Uylings (1977) gives a pressure drop that depends linearly on $Q_m l_m$ and is proportional to

$$d_m^{-4-k}$$

where k runs from 0, for laminar flow, to 1 for turbulent flow. This leads to the result

$$R_d = R^{2/(6+k)} \qquad (15.10)$$

On the other hand, the treatment by Pedley *et al.* (1970) suggests that the pressure drop is dependent on the ratio d_m/l_m, which complicates the analysis because the volume and pressure drop terms no longer depend on l_m in the same way.

For blood there is evidence that the viscosity is a function of vessel diameter—the Fahraeus–Lindquist effect. Iberall (1967) parametrizes this by replacing d_m^{-4} by

$$f(d_m) = d_m^{-2}(d_m + 6)^{-2} \qquad (15.11)$$

with d_m measured in microns. Now the smallest vessels considered in Iberall's compilation of data have d_m comparable to 6 μm, so this correction is not a very significant one.

The consequences of equation (15.9) for cross-sectional area in successive orders are readily seen. If $R_d = R^{1/\alpha}$ the ratio of cross-sections in successive orders is

$$R_a = R R_d^{-2} = R^{1-2/\alpha} \qquad (15.12)$$

For laminar flow this gives

$$R_a = R^{\frac{1}{3}}$$

The corrections implied by equations (15.10) or (15.11) reduce R_a. So they are irrelevant to the substantial increase in R_a noted by Iberall (1967) for arteries of diameter below about 20 microns. The total volume of the airways can now be estimated, using equation (15.9) and the result of the previous chapter

$$R_l = R^{\frac{1}{3}}$$

If the trachea has volume V_M and is the (unique) branch of Strahler order M, the total volume is simply

$$V = V_M R^{(M-m)} R_d^{(m-M)} R_l^{(m-M)} = M\ V_M \qquad (15.13)$$

That is, the volume of all branches of a given Strahler order m is independent of m, since the cross-section goes up as $R^{\frac{1}{3}}$ but the length goes down as $R^{-\frac{1}{3}}$.

Estimating human tracheal volume as 2.5×10^{-5} m^3 (von Hayek 1960) and taking seventeen Strahler orders of branching, gives the airway volume as 4.3×10^{-4} m^3, compared with total lung volume 6×10^{-3} m^3 (Weibel 1963). It may be noted that alveoli are bulky objects.

15.2 BRANCHING ANGLES

Empirical rules for arterial branching have been known since the last century (Roux 1895). Similar rules prevail in lung airways (von Hayek 1960). If an artery bifurcates into two branches which are equal in diameter, these make equal angles with the line of the original artery. If an artery gives off a very fine artery, this will be at about 90° from the original artery. If an artery branches into two with different diameters, the smaller of these makes the larger angle with the line of the original artery. The style of argument used so far, in which a suitable cost function is set up and minimized, can be extended to discuss the most favoured values for branching angles in arteries and airways.

Murray (1926b) developed an argument concerning the angles of bifurcation of arteries, assuming laminar flow and using a cost function containing a volume term and a resistance term. His treatment has been extended by various authors; for example, Uylings (1977) discusses non-laminar flow. The most interesting extension is that of Zamir (1976b) who examines a number of cost functions. He assumes that an arterial junction has an optimal configuration when it minimizes F, where F is one of the following:

the total lumen surface, $F = S$;
the total lumen volume, $F = V$;
the power needed to pump blood through the junction, $F = W$;
the shear force acting on the lumen walls, $F = T$. [See Appendix 3.]

It is clear that one cannot absolutely prefer one of these over the others, and a true cost function may be some combination of all of them. Thus even if the different choices of F lead to strikingly different predictions about the branching angles, this would not tell us very much. Perhaps the best to be hoped for is that the predictions should not be very sensitive to the choice of F, so that it is possible jointly to satisfy them all. The most interesting result of Zamir's work is, in fact, of this kind.

Using $F = S$ means seeking the minimum amount of arterial wall material, on the assumption of uniform thickness. Using $F = V$ means seeking the minimum volume of blood in the arterial system. Using $F = W$ means seeking the minimum rate of working of the cardiac muscles. Using $F = T$ is favoured by Zamir (1976a) on the grounds that this is more likely than the others to have a local effect at an individual junction. He regards the shear force between the blood and the tissue of the walls as a property of which the wall 'is directly and tangibly aware'. He suggests that there may be a higher frequency of arterial lesions near 'incorrect' junctions, thus providing a local mechanism for natural selection. Some results of Turner (1965) seem to count against this argument of Zamir. Turner measured the radii and branching angles at about sixty arterial branching points, on major arteries. He found that the average values conformed with a relationship by minimizing a cost function, but that at any one junction there was little correlation between the angles and the radii.

Denoting by lower case letters quantities referring to unit length of a vessel, so that $S = sl$ and so on, one has

$$s = \pi d \tag{15.14a}$$

$$v = \pi d^2/4 \tag{15.14b}$$

$$w = 128\mu Q^2/\pi d^4 \tag{15.14c}$$

$$t = 32\mu Q/d^2 \tag{15.14d}$$

Here d is diameter, Q is flow and μ is viscosity. For a single vessel one cannot minimize s, v, w, or t separately, but only some linear combination of either s or v and either w or t, as in the work of Wilson (1967) discussed above. But in the more complicated case of three vessels meeting at a bifurcation, it is possible to minimize any one of these cost functions, by specifying a suitable dependence of each of the two branching angles θ_1 and θ_2 (see Fig. 13.1) on the three diameters and, in the case of w or t, on the three flows. It is a more complicated matter to work with w or t than with s or v, because the flows are involved. Two possible strategies can be adopted to simplify the analysis. One can confine one's attention to the case of equal bifurcation angles, and equal radii in the minor branches

$$\theta_1 = \theta_2 = \theta, \ d_1 = d_2 = d$$

which is the case to be presented in some detail here. Alternatively, as in another paper by Zamir (1978), one can assume that the flows depend on the diameters by a simple power law

$$Q = \alpha d^{\gamma}$$

In the first case it is plausible to assume identity of flow in each minor vessel, with

$$Q_1 = Q_2 = \tfrac{1}{2}Q_0$$

In the second case one replaces equations (15.14c) and (15.14d) by

$$w = 128\mu\alpha^2 \, d^{2\gamma-4} \tag{15.15c}$$

$$t = 32\mu\alpha \, d^{\gamma-2} \tag{15.15d}$$

I shall relegate the fairly complicated details of Zamir's analysis to Appendix 4, and shall quote here his result for a symmetrical bifurcation into two minor branches at angles $\pm\theta$ with the line of the original artery, writing the cost function as $F = fl$, and taking subscript 0 to refer to the main artery, as in Fig. 13.1. The result is

$$\cos(2\theta) = \frac{f_0^2 - f_1^2 - f_2^2}{2f_1 f_2} \tag{15.16}$$

Symmetrical branching is understood to imply equal diameter and equal flow in each of the minor branches, so that with $d_1 = d_2 = d$ we have

$$F = S: \cos(2\theta) = \frac{d_0^2 - 2d^2}{2d^2} \tag{15.17a}$$

$$F = V: \cos(2\theta) = \frac{d_0^4 - 2d^4}{2d^4} \tag{15.17b}$$

$$F = W: \cos(2\theta) = \frac{8d^8 - d_0^8}{d^8} \tag{15.17c}$$

$$F = T: \cos(2\theta) = \frac{2d^4 - d_0^4}{d^4} \tag{15.17d}$$

If we define the cross-section ratio by

$$\beta = 2d^2/d_0^2$$

these results can be compressed into the forms

$$\cos(2\theta) = \beta^{-1} - 1,\ 2\beta^{-2} - 1,\ \beta^4/2 - 1,\ \beta^2/2 - 1 \tag{15.18}$$

respectively, and these are displayed in Fig. 15.1. The four different cost functions S, V, W, and T give very different dependence on the ratio β. However, near the value $\beta = 2^{\frac{1}{3}}$, which is the laminar flow value, it is possible approximately to minimize all the cost functions in the region $\theta = 40°$ to $\theta = 50°$.

Zamir comments that the available data refer mainly to large arteries, which may not constitute a typical sample, and stresses the need for data on nearly symmetrical bifurcations of small arteries. However, it is worth quoting in this context the remark of Murray (1926b) on unpublished data available to him, relating to arteries in the cat lung:

> 'for every bifurcation angle [which here means $\theta_1 + \theta_2$] less than let us say 70° there are hundreds if not thousands of cases where the angle is between 70° and 90°.'

Thurlbeck and Horsfield (1980) find $\theta_1 + \theta_2$ in the range 80° to 60° in a dog lung.

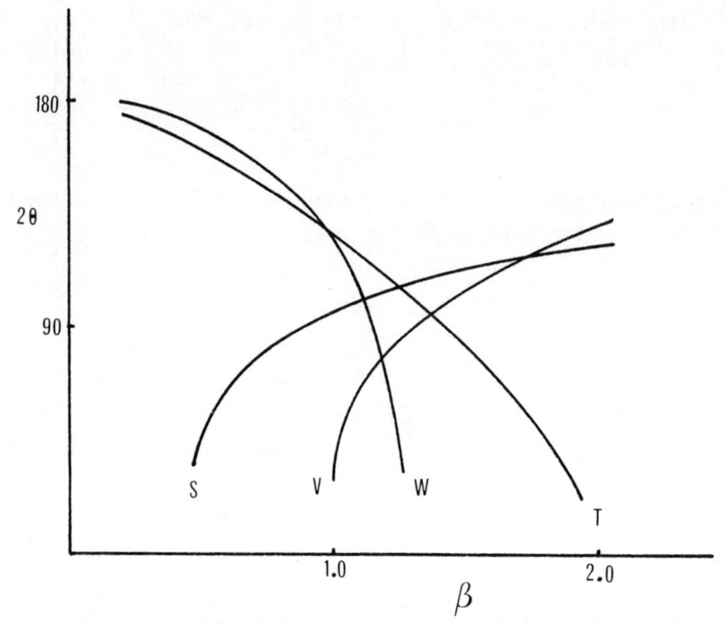

Fig. 15.1 The total branching angle 2θ for a symmetrically branching artery, plotted against the cross-section ratio β. For each curve θ(β) minimizes a cost function *S*, *V*, *W*, or *T* as indicated. Reproduced from 'Optimality principles in arterial branching', *J. theor. Biol.* **62**, 227–251, by M. Zamir, by permission of Academic Press Inc. (London) Ltd. and Dr. Zamir. © 1976 by Academic Press

Zamir (1978) analyses asymmetrical branching, using a power law for the flow

$$Q = \alpha d^{\frac{1}{3}}$$

corresponding to laminar flow, and Roy and Woldenberg (1982) extend this analysis to other power laws. The results are consistent with the laws of Roux (1895) mentioned above. Zamir *et al.* (1979) compare the results with branching of arteries in the human retina, which can be studied photographically *in vivo* (Fig. 13.6). The data do not discriminate between *S* or *W*, or between *V* or *T*.

I end this chapter by noting a basic difficulty in the analysis of branching structures, which emerges if we try to integrate global and local criteria, as in the treatments given by Wilson (1967) and by Zamir (1976b) respectively. So long as the model used has $R = 2$ one can directly compare global and local results. All branches are edges and it is simple to define the appropriate direction, length, and mean diameter for each of them. For the prevailing asymmetrical branching structures the Strahler ordering method has distinct advantages in the global analysis. However, any Strahler branch except those of order 1 is liable to have side branches (on the average one side branch if $R = 3$) and consequently to be kinked and attenuated at the point where it throws off that branch. In the symmetrical bifurcations studied by Zamir (1976), and in a model with $R = 2$,

Wilson's type of analysis, applied to arteries, leads to the value $\beta = 2^{\frac{1}{3}}$, precisely the value of particular relevance to Zamir's results. But for any other R, Wilson's argument gives $R_d = R^{\frac{1}{3}}$, which cannot be directly used to estimate β.

References

Campbell, E. J. M., Westlake, E. K., and Cherniack, R. M. (1957). A simple method of estimating oxygen consumption and efficiency of the muscles of breathing, *J. appl. Physiol.* **11**, 303–308.

Horsfield, K. (1977). Morphology of branching trees related to entropy, *Resp. Physiol.* **29**, 179–184.

Horsfield, K., and Cumming, G. (1967). Angles of branching and diameters of branches in the human bronchial tree, *Bull. math. Biophys.* **29**, 245–259.

Iberall, A. S. (1967). Anatomy and steady flow characteristics of the arterial system with an introduction to its pulsatile characteristics, *Math. Biosciences* **1**, 375–395.

Leopold, L. B., and Langbein, W. B. (1962). The concept of entropy in landscape evolution, *U.S. Geol. Survey Prof. Papers* 500A.

Murray, C. D. (1962a). The physiological principle of minimum work. I The vascular system and the cost of blood volume, *Proc. Nat. Acad. Sci. U.S.A.* **12**, 207–214.

Murray, C. D. (1926b). The physiological principle of minimum work applied to the branching of arteries, *J. gen. Physiol.* **9**, 835–841.

Pedley, T. J., Schroter, R. C., and Sudlow, M. F. (1970). Energy losses and pressure drop in models of human airways, *Resp. Physiol.* **9**, 371–386.

Rosen, R. (1967). *Optimality Principles in Biology*, Butterworth, London.

Roux, W. (1895). *Entwicklungs-mechanik der Organismen. I Funktionelle Anpassung*, Leipzig.

Roy, A. G., and Woldenberg, M. J. (1982). A generalisation of the optimal model of arterial branching, *Bull. math. Biol.* **44**, 349–360.

Thurlbeck, A., and Horsfield, K. (1980). Branching angles in the bronchial tree related to the order of the branch, *Resp. Physiol.* **41**, 173–181.

Turner, R. S. (1965). Some limitations of the theoretical approach to problems in morphology, *Anat. Rec.* **146**, 293–298.

Uylings, H. M. B. (1977). Optimisation of diameters and bifurcation ratios in lung and vascular tree structures, *Bull. math. Biol.* **39**, 509–519.

von Hayek, H. (1960). *The Human Lung*, Hafner, New York.

Weibel, E. R. (1963). *Morphometry of the Human Lung*, Springer, Berlin.

Wilson, T. A. (1967). Design of the bronchial tree, *Nature* **213**, 668–669.

Zamir, M. (1976a). The role of shear forces in arterial branching, *J. gen. Physiol.* **67**, 213–222.

Zamir, M. (1976b). Optimality principles in arterial branching, *J. theor. Biol.* **62**, 227–251.

Zamir, M. (1978). Nonsymmetrical bifurcations in arterial branching, *J. gen. Physiol.* **72**, 837–845.

Zamir, M., Medeiros, J. A., and Cunningham, T. K. (1979). Arterial bifurcations in the human retina, *J. gen. Physiol.* **74**, 537–548.

Chapter 16

Pipes, bundles, and horns: more biophysics

This chapter deals with models in which branching angles are ignored. A branching structure is discussed as if it were a single trunk, or as if the branches, although distinct and each of a specific length, were aligned in a bundle. Although these models greatly over-simplify the structure, they are useful because they are directed at specific limited questions. The pipe model was introduced to the modern botanical literature by Shinozaki *et al.* (1964a). They do not refer to the earlier history of the concept of a tree as a bundle of pipes, terminating in the leaves, which was a commonplace of Victorian botany. John Ruskin's *Modern Painters* contains several eloquent pages on this concept, and even mentions some of its quantitative consequences.

16.1 THE PIPE MODEL FOR A BOTANICAL TREE

Shinozaki *et al.* (1964a) observe that when leaves are counted, and trunk and branches weighed, in horizontal slices across a tree, there is a correlation between these quantities. The total trunk and branch weight at a certain height varies in the same way as the total leaf number above that height. They suggest that if all branches existing at a given height support, mechanically and functionally, all the leaves above that height, then denoting the leaf count in slice dh at height h by $L(h)dh$ and the branch weight in slice dh at height h by $B(h)dh$, one should expect

$$B(h) \propto \int_h^H L(h)dh \tag{16.1}$$

where H is the total height of the tree. Fig. 16.1 shows their best example, based on data for the spruce *Picea glehni.* In general they find that deviations from equation (16.1) are most noticeable in the trunk, where the biomass and the cross-section can continue to increase right down to the ground, although there are no leaves at all at very low h. This is consistent with the loss of low-lying branches as

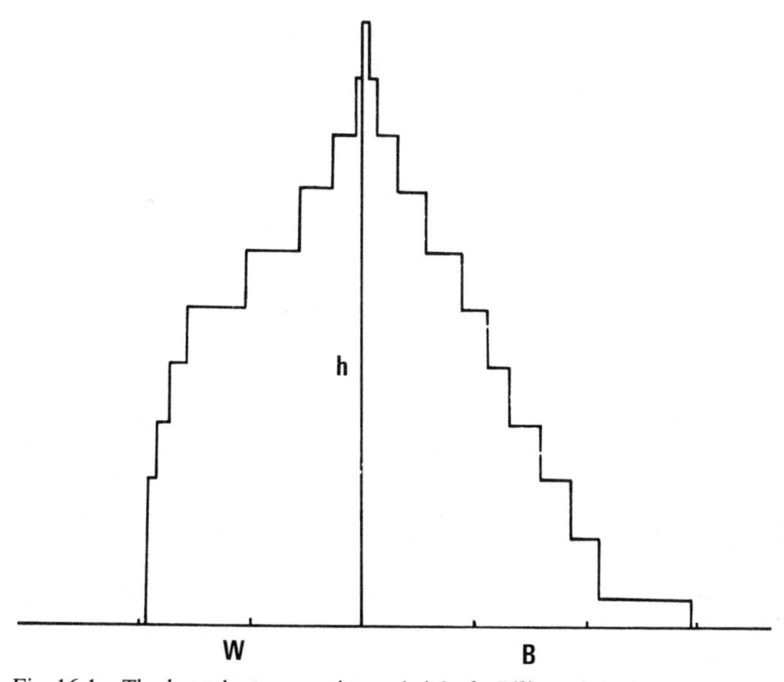

Fig. 16.1 The branch cross-section at height h, $B(h)$, and the integrated leaf weight at heights greater than h, $W(h) = \int_h^H L(h)dh$, plotted against h. The data are for the spruce *Picea Glehni*. Reproduced from 'A quantitative analysis of plant form—the pipe model theory I', *Jap. J. Ecol.*, **14**, 97–105, by K. Shinozaki *et al.*, by permission of the Ecological Society of Japan. © 1964 by the Ecological Society of Japan

the tree matures, particularly in close stands where these branches receive little light.

The pipe model is a description of the tree consistent with equation (16.1). The trunk is taken to be made up of a large number of identical pipes, of a fixed cross-section. Each of these ultimately leads, through various successive branches, to a single leaf. The pipe acts as a mechanical support and as a vascular passage, and extends from the leaf to the base of the trunk. Any branch is made up of a bundle of pipes. The cross-section of the branch is proportional to the number of leaves which it supports directly or indirectly. If one assumes that the leaves are distributed evenly over the branches of Strahler order 1, then the cross-section of a branch of Strahler order m is proportional to R^m. Consequently in this model

$$R_d = R^{\frac{1}{2}}$$

Another way of seeing this is that all leaves are supported (indirectly) by all branches of any order, so that the total cross-section of branches of order m must be independent of m. Taking a local point of view instead, and considering a single bifurcation, the cross-section of the parent stem must equal the sum of the cross-sections of the two daughter branches. In the notation of Fig. 13.1

$$d_0^2 = d_1^2 + d_2^2$$

This and further consequences of the pipe model, already outlined in Chapter 13, are discussed in the second paper of Shinozaki *et al.* (1964b), along with an extension to the root system. Fig. 16.2 illustrates the pipe model concept, including the effects of lost branches.

While McMahon and Kronauer (1976) present some results on the Horton branching ratio for a few botanical trees, included in Table 13.1, their work is essentially concerned with a bundle model. The fact that Horton's law seems to apply to trees encouraged them to seek evidence relating to other self-similarity properties of a geometrical or mechanical, rather than topological, kind. At any sample point on a branch they recorded the diameter, and also the cumulative length to the point from the base of the trunk, along branches of decreasing Strahler order. They find that this diameter d and cumulative length L are correlated, in the six trees examined, according to an approximate power law

$$d = aL^b \tag{16.2}$$

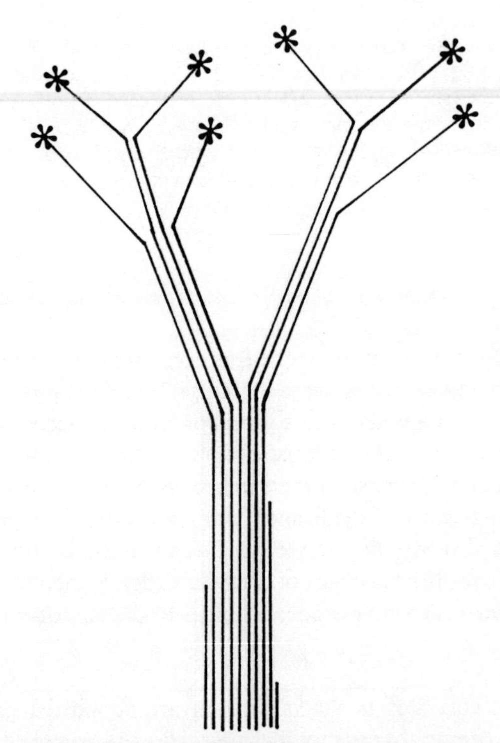

Fig. 16.2 A schematic illustration of the pipe model

with $b = 1.50 \pm 0.13$. They suggest for comparison three alternative theoretical values of b. Geometrical similarity gives $b = 1$, elastic similarity $b = 1.5$, and stress similarity $b = 2.0$. Here elastic similarity means that the deflection of the branch, under its own weight and that of the branches it carries, and resisted by its elastic strength, is proportional to the cumulative length L. Stress similarity means that the maximum stress at any point of an arbitrary cross-section is independent of L. They conclude that, if it is indeed appropriate to assign a single mechanical 'design criterion' to these trees, elastic similarity is an appropriate one.

Niklas (1978a) analyses his data on fossil plants (Niklas 1978b) in this way, finding that the highly symmetrical Rhyniophytes show very little tapering, with values of β ranging from 0 to 0.7. The Zosterophyllophytes have β near 1.5, and the Progymnosperms have β ranging from 1.8 to 1.9.

16.2 RALL'S MODEL OF DENDRITES

Rall (1959, 1962) developed a model of the motoneurons of the cat in which the dendritic tree is treated as a single cable, and linear cable theory is used. It is generally believed that in these cells no action potential develops on the dendrites. This means that linear cable theory is adequate, and that signals attenuate as they pass along a dendrite. Rall's aim was to assess the relative importance of signals from synapses situated near to and far from the cell body. Even linear cable theory would be hard to apply to a branching structure of cables, and a pipe model yields substantial simplification. Rall's model leads to a relationship between the diameters of branches at a junction. If the model is to be used it is necessary to check that this relationship between diameters holds to a fair approximation.

To see how the model leads to a specific conclusion about diameters, it is first necessary to put the linear cable equation into dimensionless form, and to indicate how various electrical parameters depend on cable diameter. Since nerve cells have negligible inductive properties the cable equation reduces to a quasi-diffusive form (see, for example, Hallén (1962)). This is

$$\frac{\partial^2 V}{\partial x^2} - Cr\frac{\partial V}{\partial t} + rGV = 0 \tag{16.3}$$

Here V is the potential difference across the cell membrane, x is the distance along the cylindrical dendrite, and t is time. The capacitance per unit length of the dendrite is $C = \pi dc$, where d is the diameter and c is the capacitance of unit area of membrane, a quantity that is independent of d. The shunt conductance per unit length of membrane is $G = \pi dg$, where g is the conductance per unit area of membrane, again a quantity independent of d. Finally, the internal resistance of unit length of the dendrite is

$$r = \frac{4}{\pi d^2 \sigma}$$

where σ is the conductivity of a unit cube of dendritic interior, also independent of d. We now have all the information we need about the way electrical properties depend on d. (Full details of the cylindrical cable can be found in Scott (1977).)

Equation (16.3) is reduced to a dimensionless form by scaling to new variables

$$X = (rG)^{\frac{1}{2}}x \qquad T = Gt/C \tag{16.4}$$

and becomes

$$\frac{\partial^2 V}{\partial X^2} = V + \frac{\partial V}{\partial T} \tag{16.5}$$

The condition for a continuous unbranched pipe treatment is that V changes smoothly with X, that is $\partial V/\partial X$ is continuous. Current must also be conserved at any branch point. Let us again denote an incoming branch by the subscript 0 and two outgoing branches by subscripts 1 and 2. The current conservation condition is

$$\left.\frac{\partial V}{\partial x}\right|_0 \frac{1}{r_0} = \left.\frac{\partial V}{\partial x}\right|_1 \frac{1}{r_1} + \left.\frac{\partial V}{\partial x}\right|_2 \frac{1}{r_2} \tag{16.6}$$

In the dimensionless variables this is

$$\left.\frac{\partial V}{\partial X}\right|_0 \left(\frac{G_0}{r_0}\right)^{\frac{1}{2}} = \left.\frac{\partial V}{\partial X}\right|_1 \left(\frac{G_1}{r_1}\right)^{\frac{1}{2}} + \left.\frac{\partial V}{\partial X}\right|_2 \left(\frac{G_2}{r_2}\right)^{\frac{1}{2}}$$

which by the results in the last paragraph is

$$\left.\frac{\partial V}{\partial X}\right|_0 d_0^{3/2} = \left.\frac{\partial V}{\partial X}\right|_1 d_1^{3/2} + \left.\frac{\partial V}{\partial X}\right| d_2^{3/2} \tag{16.7}$$

To reconcile equation (16.7) with continuity of the gradient of V

$$\left.\frac{\partial V}{\partial X}\right|_0 = \left.\frac{\partial V}{\partial X}\right|_1 = \left.\frac{\partial V}{\partial X}\right|_2 \tag{16.8}$$

it is necessary to have the relationship between the three diameters

$$d_0^{3/2} = d_1^{3/2} + d_2^{3/2} \tag{16.9}$$

As compared with the Shinozaki pipe model, in which the cross-section is constant, and the arterial or bronchial trees, in which the cross-section increases outwards, the consequence of equation (16.9) is that the pipe cross-section shrinks as one goes outwards from the cell body. For example, in a symmetrical branching model the areas in successive orders fall off as $2^{-\frac{1}{3}}$. Dendrite diameters are difficult to measure because of surface irregularities. However, Lux *et al.* (1970) present data on 50 bifurcations in seven cat motoneurons for which the ratio of $d_0^{3/2}$ to

$d_1^{3/2} + d_2^{3/2}$ is 1.02 ± 0.12. Barrett and Crill (1971) study the tapering of dendrites and conclude that it occurs at bifurcations according to equation (16.9) but that there is additional tapering along the length of each segment.

16.3 MODELS FOR IMPEDANCE CALCULATIONS WITH ARTERIES AND AIRWAYS

The flow of blood in arteries is not in fact steady, as assumed in Chapter 15, but involves a wave-like pulse initiated by the heartbeat. If all the arteries together behaved like one uniform tube, with a reflecting wall at the end, this pulse would be reflected back to the heart. Naturally the observed response is much more complicated. It is usual to analyse the response in terms of incident harmonic waves, since any pulse can be treated as a superposition of such waves. If the response is dependent on the frequency of the incident harmonic wave, then the pulse will not be necessarily reflected in a clearly recognizable form. The relationship between incident and reflected harmonic waves can be characterized by impedance, as described in Appendix 5.

Taylor (1966) calculates the impedance of a symmetrically branching arterial system, with the vessels of highest (Weibel) order terminated by identical reflecting walls. Although he illustrates his work with diagrams of perpendicularly bifurcating trees, he relies on the fact that for wavelengths sufficiently long compared with vessel diameters the angles of bifurcation are irrelevant to reflection at the bifurcation. As discussed in Appendix 5, in these circumstances reflection at a bifurcation depends only on the relative values of the cross-sectional areas. It is also shown in the Appendix that the bifurcating daughter vessels are treated as parallel in the sense used in electrical circuit theory, that is to say one combines their impedances by adding their reciprocals. So in a slightly punning sense one can say that this is a bundle model. In Taylor's model the nature of the reflected pulse depends on reflection at many bifurcations and on the phase relations caused by the spread of values of the total path from the heart to a reflecting wall and back.

Taylor does not use a power law for the lengths of successive branches, as described in Chapters 13 and 14. Instead he uses

$$\bar{l}_m = \frac{1}{m+1} l_0$$

where there is one branch of order 0 and ordering is centrifugal. As m increases, the ratio of $\bar{l}_m$ to $\bar{l}_{m-1}$ tends to 1. For each order m he takes lengths l distributed around $\bar{l}_m$ in a gamma distribution. He takes m up to seven, a considerable simplification compared with the real system. His method, as discussed in Appendix 5, is to work inwards from the impedances of the identical loads at the ends of the branches of order 7. From these one can calculate impedances of effective loads at the ends of branches of order 6, and so on until one has the impedance of the whole tree. Taylor finds that the reflection properties are dominated by a single low-frequency resonance.

The inference of lung structure from the response of the lung to an input sound pulse goes back to the eighteenth century adoption of the diagnostic trick of listening to the sound produced when the chest is tapped by the fingers. More sophisticated techniques employ harmonic wave inputs to the trachea, and in recent years these have spanned a range of frequencies up to several thousand hertz. The results have been studied using models at various levels of detail.

We have seen in Chapter 15 that the total cross-section of the branches of Strahler order m increases towards the terminal branches as $R^{(M-m)/3}$. So an appropriate pipe model for lung airways is an exponential horn, with the cross-section increasing exponentially with distance from the throat. Ishikaza *et al.* (1976) use a model of this type, based on Weibel's symmetrically branching model with $R = 2$. Using this model they attempt a fit with their own data, which indicate strong resonances, for human lungs, in the vicinity of frequencies 650, 1400 and 2100 hertz.

Sidell and Fredberg (1978) criticize the exponential horn model on the grounds of the observed asymmetry of the bronchial tree, with $R = 3$ in the case of Strahler ordering. They also point out that the method of combining impedances in parallel can be used effectively for a rather complex tree, if one makes use of a regular model, in which the scaling of cross-sections at a bifurcation, and the scaling of branch lengths, are taken to be fixed. The model they use is that of Horsfield *et al.* (1971) described in Section 14.3. In terms of Horsfield ordering the bifurcations are of the type $m \to m, m-3$. They use 35 orders of branching, giving about 10^6 branches of order 1. Whereas in the horn model all path lengths from trachea to alveolar ducts are the same, Sidell and Fredberg have paths with lengths from 16 to 35 cm. (The regular model has to be truncated in the manner shown in Fig. 14.6, at Horsfield orders lower than 4.)

References

Barrett, J. N., and Crill, W. E. (1971). Specific membrane resistivity of dye injected cat motoneurons, *Brain Research* **28**, 556–561.

Hallén, E. (1962). *Electromagnetic Waves*, Chapman and Hall, London.

Horsfield, K., Dart, G., Olson, D. E., Filley, G. F., and Cumming, G. (1971). Models of the human bronchial tree, *J. appl. Physiol.* **31**, 207–217.

Ishikaza, K. M., Matoudaira, M., and Kaneko, T. (1976). Input acoustic impedance of the subglottal system, *J. acous. Soc. Amer.* **60**, 190–197.

Lux, H. D., Schubert, P., and Kreutzberg, G. W. (1970). Direct matching of morphological and physiological data in cat spinal motoneurons, in *Excitatory Synaptic Mechanisms* (edited by P. Andersen and J. K. S. Jansen), Oslo University Press, Oslo, pp. 189–198.

McMahon, T. A., and Kronauer, R. E. (1976). Tree structures: deducing principles of mechanical design, *J. theor. Biol.* **59**, 443–466.

Niklas, K. J. (1978a). Branching patterns and mechanical design in Palaeozoic plants: a theoretical assessment, *Ann. Bot.* **42**, 33–39.

Niklas, K. J. (1978b). Morphometric relationships and rate of evolution among Palaeozoic vascular plants, in *Evolutionary Biology* 11 (edited by M. K. Hecht), Plenum, New York, pp. 509–543.

Rall, W. (1959). Branching dendritic trees and motoneuron membrane resistivity, *Exptl. Neurol.* **1**, 491–527.

Rall, W. (1962). Theory of physiological properties of dendrites, *Ann. N.Y. Acad. Sci.* **96**, 1071–1091.

Scott, A. C. (1977). *Neurophysics*, Wiley, New York.

Shinozaki, K., Yoda, K., Hozami, K., and Kira, T. (1964a). A quantitative analysis of plant form—the pipe model theory. I Basic analysis, *Jap. J. Ecol.* **14**, 97–105.

Shinozaki, K., Yoda, K., Hozami, K., and Kira, T. (1964b). A quantitative analysis of plant form—the pipe model theory. II Further evidence of the theory and applications to forest ecology. *Jap. J. Ecol.* **14**, 133–139.

Sidell, R. S., and Fredberg, J. J. (1978). Non-invasive inference of airway network geometry from broadband lung reflection data, *Trans. A.S.M.E., J. biomech. Eng.* **100**, 131–138.

Taylor, M. G. (1966). The input impedance of an assembly of randomly branching elastic tubes, *Biophys. J.* **6**, 29–51.

Chapter 17

Simulating the growth of dendritic trees

Simulation can merely mean copying, that is generating on a computer a branching structure resembling one found in nature, introducing any *ad hoc* rules needed to improve the resemblance. For example, the general appearance of a botanical tree will be influenced by the branching ratio R, the branching angles, the branch length ratio R_l and the curvature of the branches in the vertical plane. Honda (1971) presents a number of structures resembling botanical trees, generated using various assumptions about bifurcation angles and about the scaling of branch lengths, and projected onto horizontal and vertical planes. Hogeweg and Hesper (1974) discuss how many distinctly different overall configurations can be produced from various combinations of branching rules. Frijters (1978) argues that if a concise expression can be given for the rules used to simulate the form of a plant, this can both serve as an objective description of the form and aid an assessment of the amount of genetic information needed to 'program' the form. These two papers both make use of the method of L-systems, which will be discussed in the next two chapters. Here the reader will note the congruence of Frijters' viewpoint with the approach of Papentin (1980) to complexity, discussed in Chapter 5.

A more ambitious kind of simulation uses the mature form of a branching structure to test hypotheses about growth that are not immediately suggested by looking at the real structure. For example, rather than organize the simulation of a lung airway system to ensure that $R = 3$, it would be preferable to take hypotheses about local growth and branching and compare the R values arising from the various hypotheses. In the example to be discussed here in some detail, the topological features of the branches of the first two or three Strahler orders are compared in real and simulated dendritic trees employing two alternative growth hypotheses. Hollingworth and Berry (1975), whose data on R and R_l for Purkinje cells are included in Table 13.1, apply this method to Purkinje cells, and separately to the basal and optical dendrites of pyramidal cells, in the rat.

17.1 TERMINAL AND SEGMENTAL GROWTH MODELS

It is known (Morest 1970) that there are active regions, known as growth cones, both at the tips of terminal dendrites and along the length of dendrites, whether terminal or not. Hollingworth and Berry formulate two extreme hypotheses, which can be called 'segmental branching' and 'terminal branching'. In the first any segment is equally likely to throw out a branch; in the second, only terminal segments can do so.

Fig. 17.1 shows the dendritic tree of a Purkinje cell from a pigeon brain. These cells are somewhat atypical of nerve cells in that the dendritic tree lies

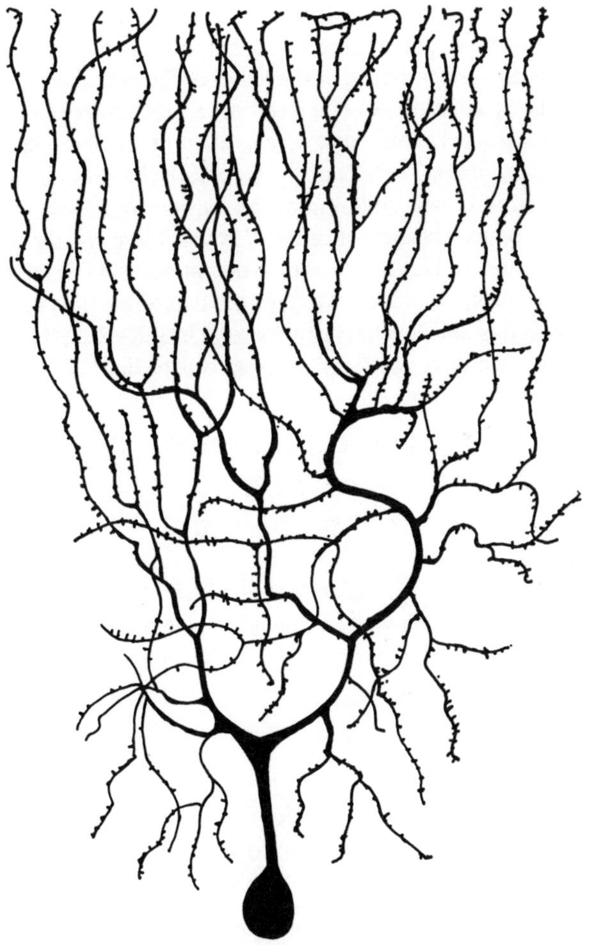

Fig. 17.1 The dendritic tree of a Purkinje cell from the cerebellum of a pigeon. Redrawn by M. F. MacDonald from Ramon y Cajal, *Histologie du Système Nerveux*, Paris, 1909

approximately in a plane. Also the end segments are not markedly longer than interior segments. (Branch lengths, however, do fall off with decreasing (Strahler) order, since the number of segments in a branch increases with order.) Besides estimating the ratio R, Hollingworth and Berry provide much information on the detailed nature of the dendritic tree in the rat's Purkinje cells. For example, in order 6 networks

$$\frac{N_1 - 2N_2}{N_1} = 0.36$$

and of the excess branches of order 1, 75% are lateral branches from branches of order 2, and 20% lateral branches from branches of order 3.

The simulation method adopted by these authors tests the models of growth against even more detailed statistics of the lower order branches. If we look at the lowest two or three Strahler orders of a tree, we can recognize the appearance of a variety of topological units. These can be classified by the number of branches of order 1 which they contain. There is only one unit of the simplest type, that is the order 1 branch, and only one unit—Y, or 2(1,1)—formed from two branches of order 1 meeting to form a branch of order 2. Again there is only one unit having three branches of order 1. This is a Y with a side arm, or 2(1)2(1,1). However, for the case of four branches of order 1 we have the two different types of unit shown in Fig. 17.2. The first is 2(1)2(1)2(1,1), the second is 3(2(1,1),2(1,1)). Going to five branches of order 1 we have the three units shown in Fig. 17.3, which are, from left to right

2(1)2(1)2(1)2(1,1)

3(2(1,1),2(1)2(1,1))

3(1)3(2(1,1),2(1,1))

The number of types of unit then increases roughly exponentially with the number of order 1 branches involved.

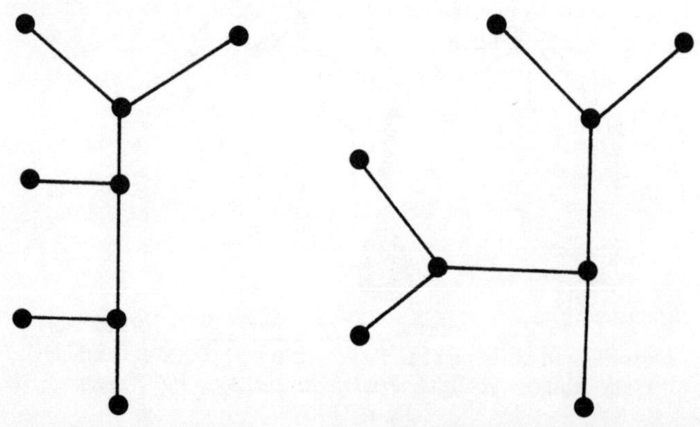

Fig. 17.2 Topological units with 4 branches of Strahler order 1

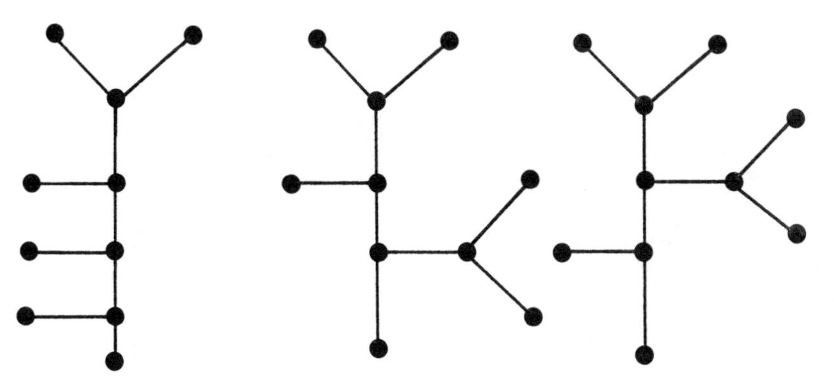

Fig. 17.3 Topological units with 5 branches of Strahler order 1

In the Purkinje cells, Hollingworth and Berry (1975) count 469 units of type 2(1)2(1)2(1) and 197 of the type 3(2(1,1),2(1,1)). The terminal growth model gives relative probabilities 2:1 and the segmental growth model gives relative probabilities 4:1 of these forms. On counting all the units with five, six or seven branches of order 1, as in Table 17.1, the comparison strongly favours the terminal growth model. Both models give all the possible types of unit but with quite different relative probabilities, as the table shows. In each row of the table the types are arranged in decreasing order of probability in the terminal growth model. It is clear that the sequence of probabilities in the real dendrites agrees more closely with the sequence in the terminal growth model than with that in the other model.

Table 17.1 also gives the analogous results for the basal dendrites of pyramidal cells. The samples studied are smaller, but a similar conclusion can be drawn.

To obtain these theoretical probabilities each model starts from a simple Y tree. In the terminal growth model, a bifurcation is assigned with equal probability to each arm of the Y, giving a figure with three tips. Again a bifurcation is assigned to each of these at random. The process continues until a tree with maximum Strahler order equal to that of the real dendritic tree is produced. A large number of these trees are generated and the relative probabilities of the different types of terminal units evaluated. In the segmental model Y is described as a set of three segments. An arm is attached to any one of these at random, giving a figure with five segments. An arm is attached to one of these at random, and so on until a suitably complex tree is obtained. Again a large number of trees are generated and the relative probabilities compiled. In a later paper Berry and Pymm (1981) give some explicit asymptotic results for these relative probabilities in large trees.

The particular interest of this work is that from the mature trees alone one can infer something about growth processes of dendritic trees. In the Purkinje cells and in the basal dendrites of pyramidal cells the terminal growth model is favoured. Employing a rather different test, Smit *et al.* (1971) also conclude that the basal dendrites of pyramidal cells, in their case from a rabbit, conform with a terminal growth model rather than with a segmental growth model. An advantage

Table 17.1 Numbers of topological units in Purkinje cells and in pyramidal cells, compared with the relative probabilities of their occurrence in the terminal growth model and the segmental growth model. N stands for the number of branches of Strahler order 1 in each unit. For each N, the first row gives the numbers of units counted in Purkinje cells, the second gives the number counted in the basal dendrites of pyramidal cells. The third row gives the relative probabilities in the terminal growth model, the fourth row those in the segmental growth model

N											
4	469	197									
	136	71									
	0.667	0.333									
	0.80	0.20									
5	203	166	66								
	106	58	30								
	0.5	0.333	0.167								
	0.286	0.571	0.143								
6	98	67	58	40	51	24					
	37	26	27	30	23	6					
	0.267	0.20	0.20	0.133	0.133	0.067					
	0.191	0.095	0.191	0.047	0.381	0.095					
7	40	42	23	32	22	25	18	11	8	10	7
	19	12	17	9	11	8	5	6	6	5	3
	0.222	0.167	0.111	0.111	0.089	0.067	0.067	0.056	0.044	0.044	0.022
	0.121	0.061	0.03	0.121	0.121	0.061	0.121	0.03	0.03	0.242	0.061

Reproduced from 'Network analysis of dendritic fields of pyramidal cells in the neocortex and Purkinje cells in the cerebellum of the rat', *Phil. Trans. Roy, Soc.* **B270**, 227–262, by T. Hollingworth and M. Berry, by permission of the Royal Society and of Prof. Berry.

of the procedure described here is that the data are taken only from the lowest Strahler orders, in which statistical regularities should be best expressed. This method might, for example, be suitable for studying the edges of mature fungal colonies, extending the work of Ho (1978).

In general one might hope to learn more from simulation if one can obtain successive pictures of the tree as it grows, but this is not often possible. It can be argued (Uylings *et al.* 1975) that if one wishes to study the sequence of development of a tree a centrifugal ordering system may be better. One approach, which has been advocated in the case of algal growth by Lück and Lück (1974), and which involves counting both Weibel (centrifugal) orders and Shreve (centripetal and cumulative) orders, will be discussed later. It seeks recurrence relations between the numbers of branches of successive orders; the motivation for this will be found in the next chapter, which deals with the L-system approach to simulation.

17.2 AN IDENTITY FOR TERMINAL GROWTH TREES

A recent result of Siegel and Sugihara (1983), on trees constructed by the terminal growth rule, suggests a possible alternative way of comparing terminal growth and segmental growth models.

Let i and j be two leaves chosen at random in a binary tree constructed according to the terminal growth rule. Let k be the outermost common vertex on the (unique) paths from the root to i and to j. Let N_i be the lengths of (number of edges on) the paths from i to k and from j to k respectively. Let $E_n^T(N_i + N_iN_j - N_i^2)$ be the mean value, in a tree with n leaves, of the quantity in brackets. Then Siegel and Sugihara show, by an inductive proof on n, that

$$E_m^T(N_i + N_iN_j - N_i^2) = 1 \tag{17.1}$$

If the corresponding quantity in a tree constructed by the segmental growth rule, $E_n^S(N_i + N_iN_j - N_j^2)$, is appreciably different from 1, evaluating the empirical mean $E_n(N_i + N_iN_j - N_j^2)$ affords a test of these models. The terminal growth model favours rather symmetrical trees, whereas segmental growth favours the appearance of long branches, such as 2(1)2(1)2(1)2(1)2(1,1) in the notation used in this chapter. The consequences of this can be seen by taking a closer look at the evaluation of E_n^T or E_n^S. When i and j are an equal distance from k

$$N_i + N_iN_j - N_i^2 = N_i \tag{17.2}$$

When i and j are at unequal distances, the mean must be calculated assuming that $N_i > N_j$ and $N_j > N_i$ are equally likely, giving a contribution

$$\begin{aligned} &\tfrac{1}{2}\{N_i + N_iN_j - N_i^2 + N_j + N_jN_i - N_j^2\} \\ &= \tfrac{1}{2}[(N_i + N_j) - (N_j - N_i)^2] \end{aligned} \tag{17.3}$$

This quantity can become negative for large enough $|N_j - N_i|$. Thus we expect E_n^S to be less than 1, and to fall as n increases, because long branches are associated with large $|N_j - N_i|$. In fact the cases $n = 4, 5, 6, 7$, and 8 give values of E_n^S equal to 0.867, 0.658, 0.396, 0.180 and –0.215 respectively.

It would be interesting to have either an explicit result for E_n^S, or an asymptotic result valid for large *n*. It would also be interesting to investigate whether some other function of N_i, N_j has an *n*-independent mean value for segmental growth trees.

Other extensions of the analysis of trees, with special reference to dendrites, are the subject of a substantial recent paper by Percheron (1982).

References

Berry, M., and Pymm, D. (1981). Analysis of neural networks, in *Neural Communication and Control* (edited by G. Szekely *et al.*), Pergamon, Oxford, pp. 155–169.

Frijters, D. (1978). Principles of simulation of inflorescence development, *Ann. Bot.* **42**, 549–560.

Ho, H. H. (1978). Hyphal branching systems in Phytophthora and other Phycomycetes, *Mycopathologia* **64**, 83–86.

Hogeweg, P., and Hesper, B. (1974). A model study in biomorphological description, *Pattern Recog.* **6**, 165–179.

Hollingworth, T., and Berry, M. (1975). Network analysis of dendritic fields of pyramidal cells in the neocortex and Purkinje cells in the cerebellum of the rat, *Phil. Trans. Roy. Soc.* **B270**, 227–264.

Honda, H. (1971). Description of the form of trees by the parameters of the tree-like body: effects of the branch angle and the branch length on the shape of the tree-like body, *J. theor. Biol.* **31**, 331–338.

Lück, J., and Lück, H. B. (1974). Production raméale et forme, *Rev. Cytol. Biol. Veg.* **37**, 313–322.

Morest, D. K. (1969). The growth of dendrites in the mammalian brain, *Z. Anat. Entwickl. Gesch.* **128**, 290–317.

Papentin, F. (1980). On order and complexity I, *J. theor. Biol.* **87**, 421–456.

Percheron, G. (1982). Principles and methods of the graph theoretical analysis of natural binary arborescences, *J. theor. Biol.* **99**, 509–552.

Siegel, A. F., and Sugihara, G. (1983). Moments of particle size distributions under sequential breakage with application to species abundance, *J. app. Prob.* Forthcoming.

Smit, G. J., Uylings, H. B. M., and Veldmaat-Wansink, L. (1971). The branching pattern in dendrites of cortical neurons, *Acta. Morphol. Scand.* **9**, 253–274.

Uylings, H. B. M., Smit, G. J., and Veltman, W. A. M. (1975). Ordering methods in quantitative analysis of the branching structure of dendritic trees, in *Physiology and Pathology of Dendrites* (edited by G. S. Kreuzberg), Raven Press, New York, pp. 247–254.

Chapter 18

The mathematics of tree simulation: L-systems

The study of simulations of trees has led to a formal theory of the generation of symbol strings by recursive application of basic transformation rules. This is known as the theory of Lindenmayer systems, after its originator (Lindenmayer 1968a,b, 1975), or L-systems for short. From the point of view of a mathematician this theory falls naturally into the context of automata theory and of the theory of formal and natural languages. The theory of L-systems has been presented from this viewpoint in great detail in two books by Vitanyi (1980) and by Herman and Rozenberg (1975). My aim here is to give a brief descriptive account that requires neither knowledge of, nor interest in, these wider contexts, and which emphasizes the biological motivation for L-systems. They are, in fact, employed to facilitate a program emphasized in the previous chapter, of testing local growth hypotheses against the global appearance either of a mature tree or of a sequence of stages in the development of a tree.

18.1 A SIMPLE TREE AND ITS STRING SEQUENCE

I shall start from a particularly simple example. Let us try to set down an algorithm for a tree in a plane satisfying a paracladial relationship between branches and the total tree. That means that the structure of any branch together with those branches it carries, at some time t_2, repeats the structure of the whole tree at some earlier time t_1. The structure is given at successive (equal) time intervals, and growth is by the addition of finite (equal) sections to a branch or bifurcating from a branch. The simulation can be carried out by means of two processes. The first is the generation of a new string of symbols by the application of a set of rules to each of the symbols in an existing string; the second is the visual interpretation of the symbols in terms of sections of a branch, points at which a branch begins, and points at which a branch ends. Let us start from the single symbol A and employ a small vocabulary of symbols, A, B, (, and). The rules of growth are

$$
\begin{aligned}
A &\rightarrow AB \\
B &\rightarrow (A)A \\
(&\rightarrow (\\
) &\rightarrow)
\end{aligned} \tag{18.1}
$$

At each stage the rules must be applied to every symbol in the current string. The sequence of strings is

$$
\begin{aligned}
&A \\
&AB \\
&AB(A)A \\
&AB(A)A(AB)AB \\
&AB(A)A(AB)AB(AB(A)A)AB(A)A \\
&AB(A)A(AB)AB(AB(A)A)AB(A)A(AB(A)A(AB)AB)AB(A)A(AB)AB
\end{aligned} \tag{18.2}
$$

and so on. Any substring that appears between a pair of brackets is exactly the same as some earlier complete string. Now let us interpret the symbols:

A means a step forward;
B means a step forward;
(means that a branch begins, and goes off at 90° to the right;
) means that a branch ends, and that the next A continues in the direction of the step immediately preceding the beginning of that branch.

The rules of growth do not depend on the context in which a symbol occurs, putting this example into the simplest category of L-systems. However, the rules of interpretation do refer to context; 'forward' for example is only meaningful when we know how many (lie between the A or B in question and the last A or B that lay on the main stem. The sequence of pictures given in Fig. 18.1 corresponds to the six strings of symbols in expression (18.2). It is more apparent from the pictures than from the strings that at stage 6 the seven branches off the main stem reproduce the tree as it looked at stages 4, 3, 2 (twice) and 1 (three times).

The separation of the rules into string generation rules and interpretation rules means that one can add geometrical verisimilitude (by using random step length, random orientation of branches, and so on) in the second stage to a basic topological growth pattern defined by the first stage. The symbol string manipulations can be carried out using operations, such as the concatenation of strings and the identification of the *n*th symbol in a string, that are available in standard computer languages. The symbol string manipulations can be studied as mathematical entities, so that one can obtain theorems which, for example, relate possible types of growth (in the number of symbols making up a string) to different types of rewriting rule (such as context-free or context-sensitive). These distinctions in the type of rewriting rule can have biological meaning.

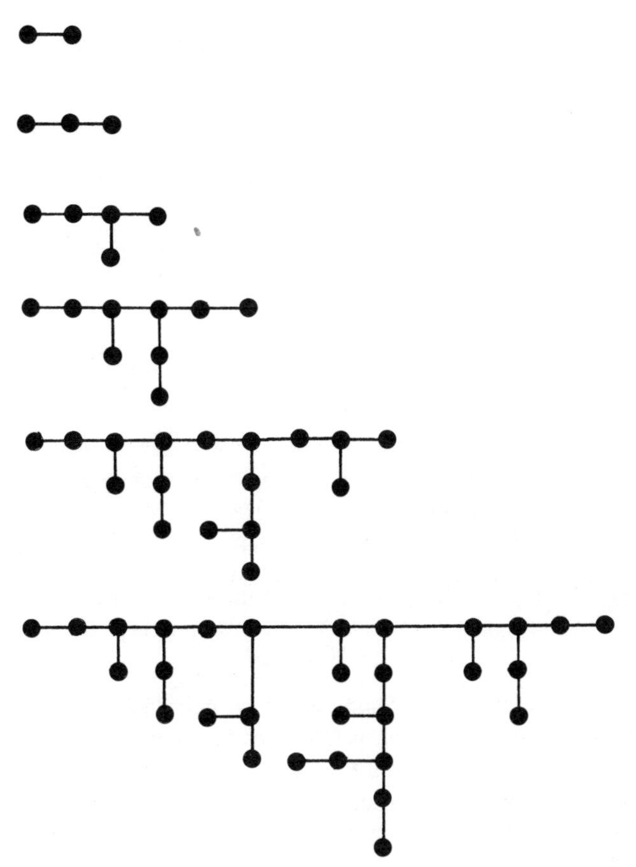

Fig. 18.1 A simple tree generated by a context-free L-system

18.2 CONTEXT-FREE AND CONTEXT-SENSITIVE L-SYSTEMS

If we think of the branching organism as made up of linear arrays of cells, like some kinds of branching algae, we may think of the symbols A and B of our example as standing for a cell in one or the other of two developmental stages, of which only the second can be followed by cell branching. Then a context-free rule means that only the current state of a cell determines the next state of that cell. A context-sensitive rule implies that some type of interaction between cells allows the states of neighbouring cells to affect the change to the next state of any particular cell. It is clearly of some interest to determine whether different types of growth can reveal anything about the need for cell interactions in the development of the branching structure. Simulations of filamentous branching algae will be discussed in the next chapter. Here I shall proceed to define first a context-free L-system, as a generalization of the symbol rewriting stage in the simple example above.

A deterministic context-free L-system, or DOL, consists of

(1) a finite non-empty alphabet X of symbols $x, y, z, \ldots$ (in my example, A, B, and (,) make up X);
(2) a starting string ξ of symbols taken from X, such as $x, yz, \ldots$ (A in my example);
(3) a finite set P of production rules, such as $x \rightarrow \eta$, where x is a symbol of the alphabet X and η is a string of these symbols or the null string, which may be written as ϕ. So $x \rightarrow x$, $x \rightarrow xyz$, $z \rightarrow \phi$ are all possible rules. However, only one rule is given for each symbol (A → AB, B → (A)A, (→ (, and) →) in my example);
(4) the meta-rule that at each step the appropriate production rule is applied to each symbol of the current string (parallel rewriting).

A probabilistic context-free L-system, or POL, has the same features, except that for one or more of the symbols there is a set of rules, such as $x \rightarrow \eta$, $x \rightarrow \zeta$, which are applied with specified probabilities, for example '$x \rightarrow \phi$ is half as likely as $x \rightarrow yz$'.

A deterministic (m, n) L-system contains the alphabet X and the starting string as before. The rules are now of the form

$$x \rightarrow \eta\,(\zeta_m, \zeta_n)$$

where ζ_m is the string of m symbols immediately preceding x, and ζ_n is the string of n symbols immediately following x. (Near the beginning or the end of the string ζ_m and ζ_n consist of up to m or up to n symbols as appropriate.) These strings form the (m, n) context of x. Again one may define a probabilistic (m, n) L-system. As an example of an L-system with a (1,1) context, consider this slight modification of expression (18.1)

A → AB unless A is preceded by (and followed by)

A → C if A is preceded by (and followed by) (18.3)

C → AB

The visual interpretation is as before, with the necessary addition that C is interpreted as a step forward. There is now a delay in the growth of all branches, and the first six stages are

A

AB

AB(A)A

AB(A)A(C)AB (18.4)

AB(A)A(C)AB(AB)AB(A)A

AB(A)A(C)AB(AB)AB(A)A(AB(A)A)AB(A)A(C)AB

which are shown in Fig. 18.2. As in Fig. 18.1 the sixth stage has thirteen steps

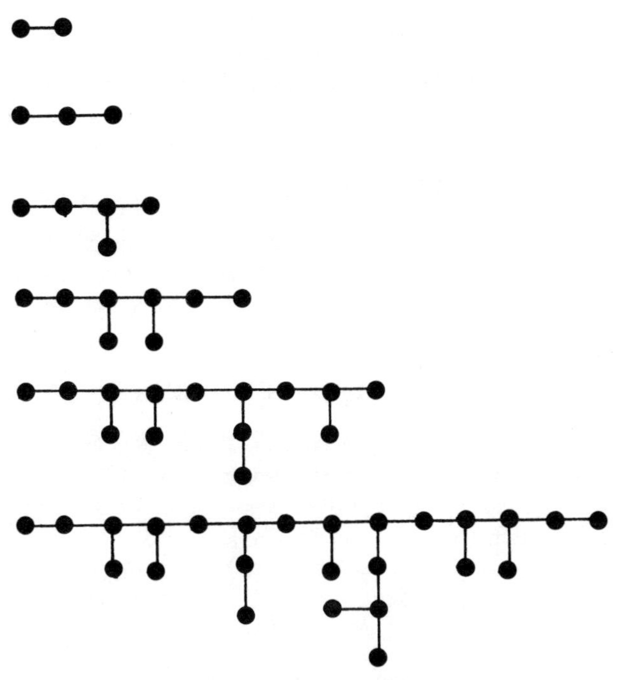

Fig. 18.2 A modified tree developed from that shown in the previous figure, by making one rule of the L-system context-sensitive, so that growth is delayed within any side branch

along the main stem, and seven branches off the main stem, but there are only a total of nine steps instead of nineteen in the branches.

It is sometimes necessary to describe two or more successive phases of growth by means of a change of the production rules. A DOL of this kind would have, for example

$$x \rightarrow \eta_1 \text{ at growth stages 1 to 10}$$

$$x \rightarrow \eta_2 \text{ at growth stages 11 and after}$$

Such a DOL is called a table DOL. This allows, for example, a phase of rapid growth followed by a slower one or one in which growth comes to a halt. As a simple example imagine that, from stage 6 on, the rules of expression (18.1) are replaced by

$$\begin{aligned} &A \rightarrow CC \\ &B \rightarrow D \\ &C \rightarrow D \\ &D \rightarrow D \\ &(\rightarrow (\\ &) \rightarrow) \end{aligned} \tag{18.5}$$

In this version stage 6 becomes

CCD(CC)CC(CCD)CCD(CCD(CC)CC)CCD(CC)CC

and all subsequent stages are

DDD(DD)DD(DDD)DDD(DDD(DD)DD)DDD(DD)DD

First the production of new branches ceases, then all growth ceases. Table DOL systems can be seen as a means of introducing effects of the environment into the otherwise self-contained development described by an L-system.

18.3 RECURRENCE RULES AND GROWTH FUNCTIONS

Now that we have defined the main types of L-system, it is useful to return to the sequence of strings of the first example (expression (18.2)) and express their relationships concisely. Inspection of these strings reveals that they can be written as a recurrence system (Herman and Rozenberg 1975)

$$\begin{aligned} S_1 &= \mathrm{A} \\ S_2 &= \mathrm{AB} \\ S_r &= S_{r-1}(S_{r-2})S_{r-2}, \; r \geqslant 3 \end{aligned} \tag{18.6}$$

These authors have shown that any sequence of strings that can be obtained by a finite set of initial string definitions and a finite set of recurrence relations can be generated by a DOL, and that any sequence of strings that is generated by a DOL can be expressed in this way. Here the recurrence relations express any string, except the initial set, in terms of previous strings and fixed strings. (The brackets in equation (18.6) are fixed strings.) In the DOL one repeatedly makes the same substitutes $[x \to \eta]$ for all symbols in a changing set of strings. In the recurrence one repeatedly substitutes different strings into the same string. The recurrence sequence expresses a global property, obtained by the application of local rules. (Here the local rules are those of expression (18.1). The global property is that each tree consists of the previous one attached to two copies of its predecessor. One of these copies forms a new major branch, while the other is an extension of the main branch.)

This result implies that, if one should be so fortunate as to encounter a real branching structure for which growth can be described by a recurrence sequence, one can model its local growth rules without employing cell–cell interactions.

In the spirit of the first few chapters of this book, one would like to find a concise way of expressing some quantitative property of the sequence of strings. One possibility is to set down the sequence of string lengths, that is to say the number of symbols in the string, and attempt to express it as a function of the number of times t that the rules have been applied. We denote the starting string again as ξ and denote by $P\eta$ the string produced by application of the production rules to each symbol in the string η. In our first example

$$\begin{aligned} \xi = P^0\xi &= \text{A} \\ P\xi &= \text{AB} \\ P^2\xi &= \text{AB(A)A} \end{aligned} \tag{18.7}$$

and so on. We denote by $L(P^t\xi)$ the number of symbols in the string $P^t\xi$. The growth function is a function from the non-negative integers into the non-negative integers defined by

$$f_G(t) = L(P^t\xi) \tag{18.8}$$

It may be possible to express $f_G(t)$ explicitly as a polynomial, or to give a linear recursion formula, that is to say a difference equation, for it, which in turn implies exponential growth. In view of the results just mentioned about the recursive properties of strings generated by DOLs, it should not be surprising that DOLs can lead to exponential growth. Let us look at two examples

1. Alphabet A, B: starting string A: rules A → B B → AB

The sequence of strings obtained is A,AB,BAB,ABBAB, BABABBAB, . . . Since A and B each yield B, the number of Bs is the number of symbols one step back. Since only B yields A, the number of As equals the number of symbols two steps back. So we have a recurrence relation

$$f_G(t + 2) = f_G(t) + f_G(t + 1) \tag{18.9}$$

and f_G is the famous Fibonacci sequence 1,1,2,3,5,8,13, . . . , which can be written explicitly as

$$f_G(t) = 5^{-\frac{1}{2}}2^{-t+1}\left[(1 + \sqrt{5})^{t+1} - (1 - \sqrt{5})^{t+1}\right] \tag{18.10}$$

2. Our original example, with production rules given by expression (18.1)

Here it is convenient to write

$$f_G(t) = f_{\text{A,B}}(t) + f_{(,)}(t) \tag{18.11}$$

where the first part is a growth function for the number of As and Bs, and the second part is a growth function for the number of brackets. From the discussion of the recurrence relation between the strings, it is clear that

$$f_{\text{A,B}}(t) = f_{\text{A,B}}(t - 1) + 2f_{\text{A,B}}(t - 2) \tag{18.12}$$

for $t = 3$ onwards, while

$$f_{\text{A,B}}(1) = 1, f_{\text{A,B}}(2) = 2$$

Also it is clear that

$$f_{(,)}(t) = 2 + f_{(,)}(t - 1) + 2f_{(,)}(t - 2) \tag{18.13}$$

for $t = 3$ onwards, while

$$f_{(,)}(1) = 0, f_{(,)}(2) = 0$$

In view of the interpretation of A and B as steps along a branch, and of (and) as the beginning and end of a branch, it seems appropriate to characterize the growth of the branching structures by either $f_{A,B}(t)$ or $f_{(,)}(t)$, depending on whether we emphasize biomass or topological complexity. Equation (18.12), with the two initial values given, has the simple solution

$$f_{A,B}(t) = 2^t \tag{18.14}$$

Equation (8.13) with the two initial values quoted has the solution

$$f_{(,)}(t) = \tfrac{1}{3}[2^t - (-1)^t] - 1 \tag{18.15}$$

The results of equations (18.10), (18.14) and (18.15) are explained in Appendix 2.

It has been proved (Vitanyi 1980) that all DOL systems have growth functions which fall into one of these four categories:

Terminating: $f_G(t) = 0$ for all $t > T$
Limiting: $f_G(t) < m$ for all t, with m an integer
Polynomial
Exponential, including cases where products of polynomial factors with exponential factors occur.

Certain examples of growth functions not in these categories, and corresponding to specific context-sensitive L-systems, have been given. (See references in the book by Vitanyi (1980).) It is, however, difficult to imagine that any data on the growth rate of a biological branching structure could be used to eliminate a DOL model as a possibility. We have little reason, for example, to associate the unit steps of t with equal intervals in real time. It is, however, intriguing to examine, as we shall in the next chapter, the possibility of detecting recurrences in the growth of branching structures.

As one goes towards more complicated types of L-system, there are fewer possibilities of applying theorems to the particular simulations, and the method seems to be mainly a means of keeping a common viewpoint and vocabulary over a range of applications. There have been many detailed simulations of branching structures which have not employed the L-system formulation, and examples can be found in the work of Bell (1976), Bell *et al.* (1979), Fisher and Honda (1977, 1979) and Honda *et al.* (1981). These could be set up in the L-system style, although little would be gained by doing this. Bell *et al.* are concerned with the growth of rhizomes which die off at the back as they grow and branch forwards; the use of the null string ϕ allows this.

It is easy to find an example of a simulation of a tree which takes a form quite different from an L-system, that by Hollingworth and Berry (1975) described in the previous chapter. Consider, for example, their segmental model. Take x as the symbol for a segment and (,) as symbols for the beginning and end of a branch.

The starting string is a Y

$$x(x)x$$

The production rules are

$$(\rightarrow ($$

$$) \rightarrow)$$

$$x \rightarrow x$$

or

$$x \rightarrow x(x)x$$

The meta-rule is that at each step $x \rightarrow x$ is applied to each x except one, picked at random, and $x \rightarrow x(x)x$ is applied to that one. It is unnatural to treat this as an example of parallel rewriting.

References

Bell, A. D. (1976). Computerised vegetative mobility in rhizomatous plants, in *Automata, Languages, Development* (edited by A. Lindenmayer and G. Rozenberg), North-Holland, Amsterdam, pp. 3–14.

Bell, A. D., Roberts, D., and Smith, A. (1979). Branching patterns: the simulation of plant architecture, *J. theor. Biol.* **81**, 351–375.

Fisher, J. B., and Honda, H. (1977). Computer simulation of branching pattern as geometry in *terminalia* (combretaceae), a tropical tree, *Bot. Gaz.* **138**, 377–383.

Fisher, J. B., and Honda, H. (1979). Branch geometry and effective leaf area: a study of terminalia branching pattern, *Amer. J. Bot.* **66**, 633–644 and 645–655.

Herman, G. T., and Rozenberg, G. (1975). *Developmental Systems and Languages*, North-Holland, Amsterdam.

Hollingworth, T., and Berry, M. (1975). Network analysis of dendritic fields of pyramidal cells in the neocortex and Purkinje cells in the cerebellum of the rat, *Phil. Trans. Roy. Soc.* **B270**, 227–264.

Honda, H., Tomlinson, P. B., and Fisher, J. B. (1981). Computer simulation of branch interaction and regulation by unequal flow rates in botanical trees, *Am. J. Bot.* **68**, 569–585.

Lindenmayer, A. (1968a). Mathematical models for cellular interactions in development. I Filaments with 1-sided inputs, *J. theor. Biol.* **18**, 280–299.

Lindenmayer, A. (1968b). Mathematical models for cellular interactions in development. II Simple and branching filaments with 2-sided inputs, *J. theor. Biol.* **18**, 300–315.

Lindenmayer, A. (1975). Developmental algorithms for multicellular organisms: a survey of L-systems, *J. theor. Biol.* **54**, 3–22.

Vitanyi, P. M. B. (1980). *Lindenmayer Systems: Structure, Languages and Growth Functions*, Mathematical Centre Tracts, No. 96, Amsterdam.

Chapter 19

Applications of L-systems

The ideal organism to simulate by an L-system would have the following characteristics: growth can be followed in terms of the number and arrangement of individual cells; all cell divisions are clocked, in the sense that although not synchronous they happen only at the ends of equal time intervals; all cells are clearly labelled to indicate to which generation they belong. It would then be possible to assign a symbol to a cell, to relate changes in the state of a cell to changes of symbol

$$x \to y$$

It would also be possible to relate cell division to production rules such as

$$x \to xx$$

or

$$x \to yz$$

and to compare sequences of symbol strings with successive stages of growth cell by cell.

19.1 A FILAMENTOUS ALGA AND A BRANCHING ALGA

As an approximation to this ideal organism Lück (1975) proposed the filamentous alga *Chaetomorpha linum.* In this the cells are arranged in a line, and there is no branching. Cell division takes place at night, so that it is reasonable to treat this as a clocked division, with a time step of one day. Careful examination of the size of the cell and of the thickness of the cell membrane allows identification of the descendants of each cell. It is found that when a cell divides the apical daughter will have a life cycle of m days, and the basal daughter will have a life cycle of n days, with $m < n$. An L-system is used in which m and n take fixed values. Lück (1975) quotes a case in which a filament with 45 cells could be divided into nineteen descendants of the first basal daughter and twenty-six descendants of the first apical daughter. The nineteen cells could be divided into eight descendants of the basal daughter of the first basal daughter, and eleven descendants of the apical daughter of the first basal daughter. The twenty-six cells could be divided into ten descendants of the basal daughter of the first apical daughter, and sixteen descendants of the apical daughter of the first apical daughter. He finds that the best fit to this is with $m=3$ and $n=5$.

A corresponding simple DOL is

$$X = a,\ b,\ c,\ d,\ e \qquad (19.1)$$
$$\xi = a$$

Production rules

$$\begin{aligned} a &\to b \\ b &\to c \\ c &\to d \\ d &\to e \\ e &\to ac \end{aligned} \qquad (19.2)$$

The sequence of strings is

$$\begin{array}{cccccccc} a & b & c & d & e & ac & bd & ce \\ dac & edb & acce & bddac & ceebd & dacacce & & \end{array}$$

and so on. The recurrence scheme for this sequence is

$$S_i = S_{i-5} S_{i-3} \qquad (19.3)$$

In the twenty-sixth string, which is made up of 48 symbols, one can identify descendants of the four symbols a, c, c, and e in the eleventh string. These are 8, 12, 12, and 16 in number respectively.

Further details of models for *Chaetamorpha linum* and other linear organisms can be found in the paper of Lück and Lück (1976). However, our primary interest is in branching organisms. There are branching algae in which the detailed cell structure of the branches can be followed. An example, studied by Konrad Hawkins (1964a,b), is *Callithamnion roseum*. An example is shown in Fig. 19.1 in which the following features should be noted. There are some apical cells which do not carry branches, and on each branch there are basal cells which do not carry branches. The branches leave the main stem alternately on either side. Some basal branches do not branch. Where branching occurs the cell walls are oblique. Konrad Hawkins (1968b) examines the lineage of cells in this organism, and Lindenmayer (1971) gives a DOL simulation

$$X = a,\ b,\ c,\ d,\ e,\ f,\ g,\ h,\ |,\ /,\ (,) \qquad (19.4)$$
$$\xi = a$$

Production rules

$$\begin{aligned}
&a \rightarrow b|c \\
&b \rightarrow b \\
&c \rightarrow b|d \\
&d \rightarrow e/d \\
&e \rightarrow f \\
&f \rightarrow g \\
&g \rightarrow h(a) \\
&h \rightarrow h \\
&(\rightarrow (\\
&) \rightarrow) \\
&/ \rightarrow / \\
&| \rightarrow |
\end{aligned} \tag{19.5}$$

So the rules incorporate change of cell state, cell division with straight wall, cell division with oblique wall, and cell division with branching. In a succession of symbols ... $g/f/e$ ... one interprets the oblique walls as tilted alternately to the left and to the right. In the succession of symbols ..$|h(a)/g$ one interprets the configuration as shown in Fig. 19.2. The first seven strings are

$$a \qquad b|c \qquad b|b|d \qquad b|b|e/d \qquad b|b|f/e/d$$
$$b|b|g/f/e/d \qquad b|b|h(a)/g/f/e/d$$

Subsequently there is a recurrence scheme

$$\begin{aligned}
S_m &= b\,b\,T_m \\
T_m &= h(S_{m-6})/T_{m-1}
\end{aligned} \tag{19.6}$$

The thirteenth sequence corresponds to the branching structure in Fig. 19.2. The feature of a non-branching basal part has been put in, in an *ad hoc* manner, by a choice of the first three production rules in (5). All branches eventually carry secondary branches.

Lindenmayer (1968) has given a more elaborate (context sensitive) L-system model of this organism. It is a D(1,1)L model using four, rather than eight cell states. Since it is context-sensitive it does not give a strict recurrence rule. In principle it would be possible to test for the presence of empirical recurrence, say in the numbers of cells in the individual branches, in an organism such as *Callithamnion roseum.*

19.2 RECURRENCE IN SHREVE ORDERS

Lück and Lück (1974) point out that it may be possible to pursue a rather similar strategy even when the tree does not consist of lines of cells. Here the quantitative

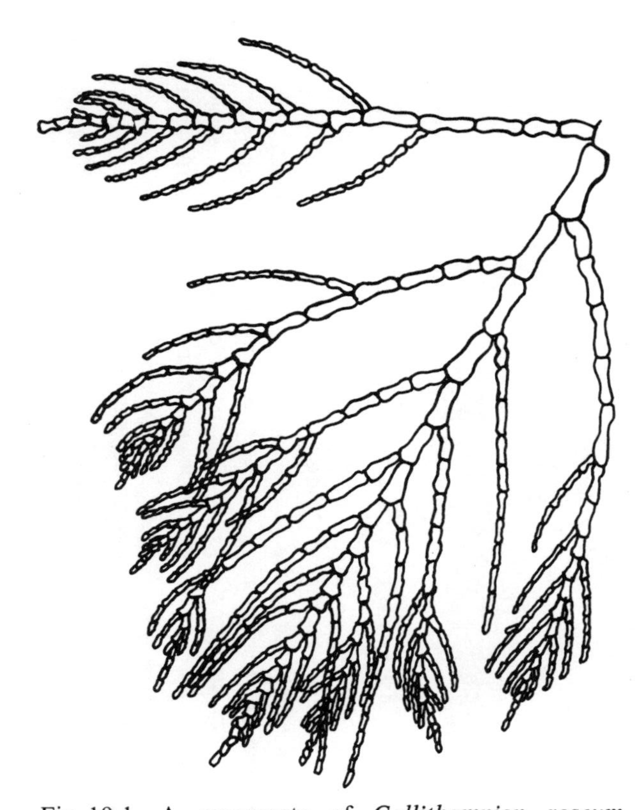

Fig. 19.1 A regenerate of *Callithamnion roseum*. Individual cells can be seen. Redrawn by M. F. MacDonald from 'Developmental studies in regenerates of *Callithamnion roseum* Harvey I', Protoplasma, **58**, 42–59, by E. Konrad Hawkins, by permission of Springer Verlag. © 1964 by Springer Verlag

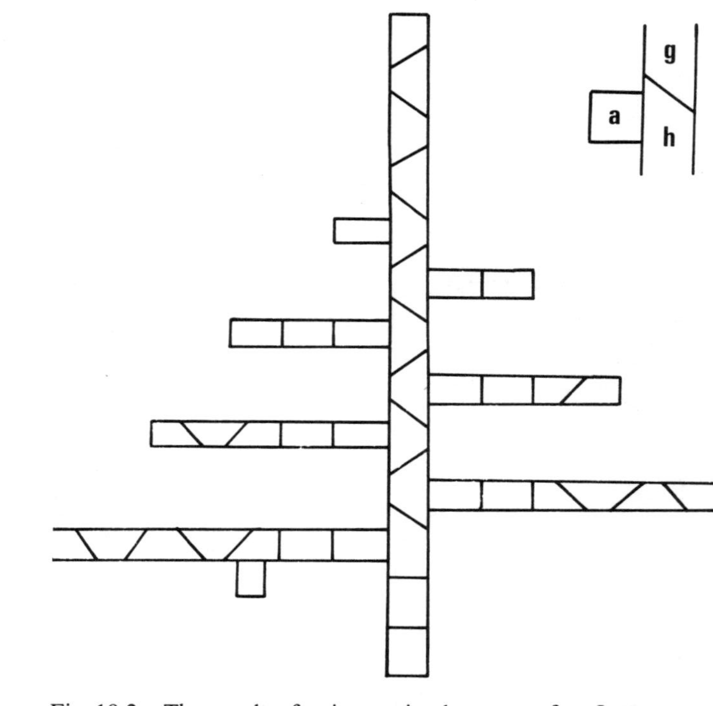

Fig. 19.2 The result of using a simple context-free L-system to simulate the cell by cell growth of regenerates of *Callithamnion roseum*. The inset shows the interpretation of the notation $h(a)/g$ in the symbol strings

data in which one seeks recurrence properties can only be the numbers of branches of Strahler order 1 subtended by the various branches of Weibel order N. One must first remove some of the randomness from the tree by drawing it as if it lay in a plane and as if at each bifurcation the branch with the larger Shreve order (that is the one which subtends the larger number of branches of Strahler order 1) always lay to the right. Fig. 19.3 shows an artificial example of a moderately asymmetrical tree ($R = 2.6$) displaying a Fibonacci recurrence relation. Successive layers of branches in the diagram are of successive Weibel orders, and the numbers appended are the Shreve orders. At Weibel order 4, for example, we see the sequence of Shreve orders

$$1\,2\,2\,3 \quad 2\,3\,3\,5 \quad 2\,3\,3\,5 \quad 3\,5\,5\,8$$

with the hierarchical structure

$$S = S_1 S_2 S_2 S_3 \tag{19.7}$$

where each subsequence has the form

$$S_i = a_i b_i b_i c_i \tag{19.8}$$

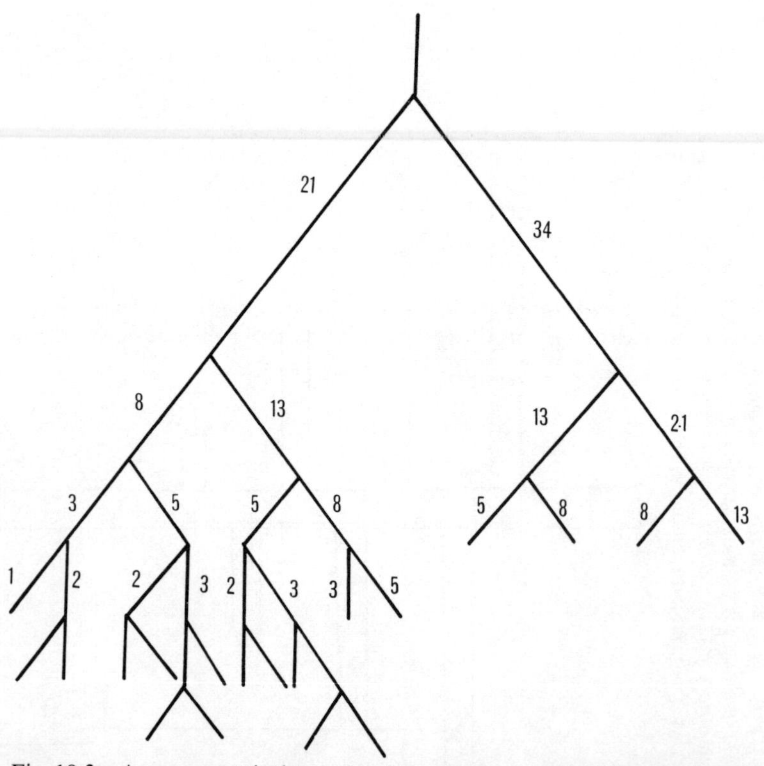

Fig. 19.3 An asymmetrical tree showing recurrence in the Shreve orders of branches which are all of the same Weibel order. Details of the low orders are omitted from the longer paths since they can be inferred from those shown

and the numbers a_1, $a_2 = b_1$, $a_3 = b_2 = c_1$, $b_3 = c_2$, and c_3 belong to the Fibonacci sequence.

If we order the branches of an empirical tree so that at this Weibel order the sequence of Shreve orders is

$$abcd \quad efgh \quad ijkl \quad mnop$$

then evidence for recurrence would consist of two features. Within each set of four numbers, the middle two lie closer to each other than to either of the outer numbers; thus

$$c - b < b - a, \; < d - c \tag{19.9}$$

A comparison of the sets of four numbers shows that the middle two lie closer to each other than to either of the outer sets.

Lück and Lück present two examples of the branching structure of the alga *Codium fragile*, extending as far as Weibel order 4. At order 2, one of these has the satisfactory sequence 6, 8, 8, 10, but the other has 6, 7, 10, 11. At order 3 the sequences are 2 4 4 4 4 4 5 5 and 2 4 3 4 4 6 4 7. In principle it would be possible to apply this test to the rather larger, and distinctly asymmetrical, trees in the data discussed in Chapter 17. I suspect that the shortcoming of this approach is not that it assumes the relevance of a context-free L-system simulation, but that it assumes the relevance of a deterministic L-system simulation.

19.3 INFLORESCENCES

As an example of the more elaborate simulations that can be carried out for branching structures when generalizations of the L-system concept are employed, I shall describe the simulations by Frijters (1978a,b) of the inflorescences of herbaceous plants, in particular of the two species *Aster novae-angliae* and *Hieracium murorum.* Inflorescences display a complex temporal and spatial pattern, starting from a single vegetative apex and culminating in a branching

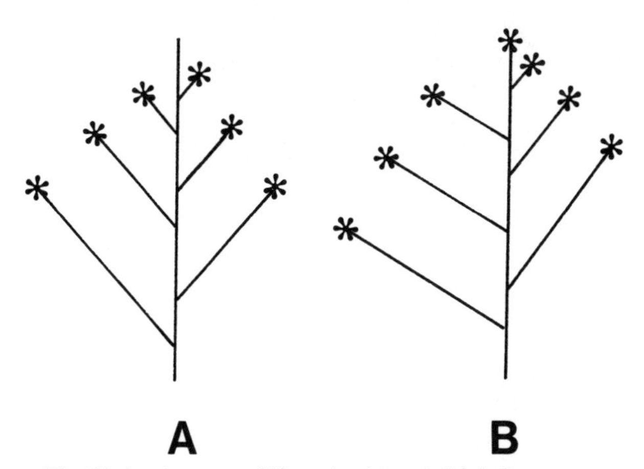

Fig. 19.4 An open (**A**) and a closed (**B**) inflorescence

structure carrying a number of flowers. As a formal example, consider the pair of similar inflorescences shown in Fig. 19.4. The first is of the type known as open, in which the apex never produces a flower, the other of the type known as closed, in which the apex as well as the branches carries a flower. In each case six branches are produced. A possible L-system for the first is

$$X = \mathrm{A,B,C,D,E,F,G,H,I,(,)}$$
$$\xi = \mathrm{A}$$

Production rules

$$\begin{aligned} &\mathrm{A} \rightarrow \mathrm{I(F)B} \\ &\mathrm{B} \rightarrow \mathrm{I(F)C} \\ &\mathrm{C} \rightarrow \mathrm{I(F)D} \\ &\mathrm{D} \rightarrow \mathrm{I(F)E} \\ &\mathrm{E} \rightarrow \mathrm{I(F)G} \\ &\mathrm{G} \rightarrow \mathrm{I(F)H} \\ &\mathrm{H} \rightarrow \mathrm{H} \\ &\mathrm{I} \rightarrow \mathrm{I} \\ &\mathrm{F} \rightarrow \mathrm{F} \\ &(\rightarrow (\\ &) \rightarrow) \end{aligned} \tag{19.10}$$

The L-system for the second differs only in having the sixth rule

$$\mathrm{G} \rightarrow \mathrm{I(F)F} \tag{19.11}$$

and omitting the seventh rule. In this simple simulation the growth of branches is ignored. The sequence (F) means a branch ending in a flower, while the symbol I means an internode and the symbols A,B,C,D,E,G,H mean successive states of the apex. Here the termination of the branching process, and the eventual flowering of the apex in the closed inflorescence, are assumed to be determined by aging of the apex. Frijters (1978a,b) also considers models in which the apex stays in one state until there is a switch to a terminating (and possibly flowering) state. This involves a table L-system, for example for the closed inflorescence

$$X = \mathrm{A,I,F,(,)}$$
$$\xi = \mathrm{A}$$

Production rules

$$\begin{aligned} &\mathrm{A} \rightarrow \mathrm{I(F)A} \\ &\mathrm{F} \rightarrow \mathrm{F} \\ &\mathrm{I} \rightarrow \mathrm{I} \\ &(\rightarrow (\\ &) \rightarrow) \end{aligned} \tag{19.12}$$

before the switch, with one change after the switch, to

$$A \rightarrow I(F)F \tag{19.13}$$

Frijters comments that an open inflorescence does not usually change into a closed inflorescence, so that one should use a model in which branching ends before flowers appear. Thus an L-system for the closed inflorescence could be

$$X = A,I,B,F,(,)$$
$$\xi = A$$

Production rules

$$\begin{aligned} A &\rightarrow I(B)A \\ B &\rightarrow B \\ I &\rightarrow I \\ (&\rightarrow (\\ ) &\rightarrow) \end{aligned} \tag{19.14}$$

before the switch, with two changes, to

$$\begin{aligned} A &\rightarrow F \\ B &\rightarrow F \end{aligned} \tag{19.15}$$

after the switch. Here A is the apex, B is a branch with no flower, I is an internode, and F is a branch or apex terminating in a flower. In this model the timing of the switch determines how many flowers are formed. In principle it should be possible to determine whether some suitable (biochemical) event occurs which constitutes the switch.

The next stage in complication is to include branches off branches, and in particular paracladia, which are branches which, in their own branching, repeat earlier stages of the whole plant. An example of such an L-system would be

$$X = A,B,C,F,I,(,)$$
$$\xi = A$$

Production rules

$$\begin{aligned} A &\rightarrow I(B)A \\ B &\rightarrow I(C)B \\ I &\rightarrow I \\ C &\rightarrow C \\ (&\rightarrow (\\ ) &\rightarrow) \end{aligned} \tag{19.16}$$

before the switch, with three changes, to

$$\begin{aligned} A &\rightarrow F \\ B &\rightarrow F \\ C &\rightarrow F \end{aligned} \tag{19.17}$$

after the switch. This gives a sequence

$$\text{A} \quad \text{I(B)A} \quad \text{I(I(C)B)I(B)A}$$
$$\text{I(I(C)I(C)B)I(I(C)B)I(B)A}$$

and if the switch takes place at that stage, finally

$$\text{I(I(F)I(F)F)I(I(F)F)I(F)F}$$

which corresponds to the inflorescence shown in Fig. 19.5. Some inflorescences include branching from the secondary branches, which can readily be added to the L-system simulation.

As well as the paracladial relationships, it may be necessary to specify the order in which flowers appear on the various branches, to characterize an inflorescence fully. In the two examples modelled in detail by Frijters (1978b) the sequences of flowering are very different. In *Hieracium murorum* flowers appear downward from the apex, while in *Aster novae-angliae* the sequence of flowering is divergent, starting with branches about one quarter of the way down from the tip, and working upwards and downwards. In the L-system simulation, a sequence of flowering times can be attained by reverting to the concept of successive stages of the apex and of each branch tip.

Quite complicated inflorescences can be simulated if one allows two generalizations (Frijters and Lindenmayer 1976) from the simple kind of L-system used in the examples above. One is to allow different rates of addition of internodes in side branches as compared with the main stem. The other is to include a delay before a

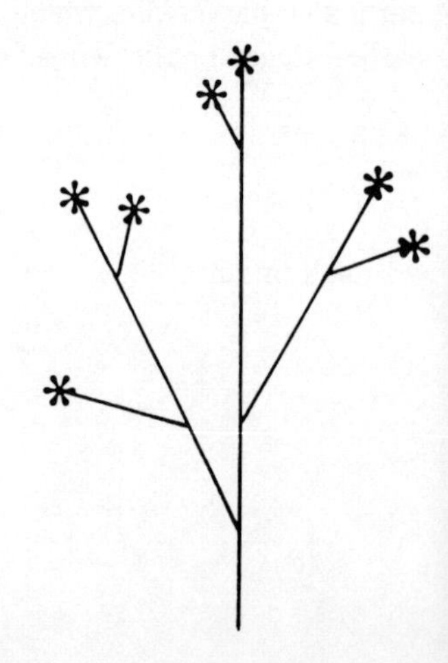

Fig. 19.5 A closed inflorescence with branches off branches

branch begins to form side branches. These authors emphasize that the requirement of different growth rates means a context-sensitive L-system. In the two detailed applications made by Frijters (1978b), *Aster novae-angliae* requires a context-sensitive model, but *Hieracium murorum* does not.

References

Frijters, D. (1978a). Principles of simulation of inflorescence development, *Ann. Bot.* **42**, 549–560.

Frijters, D. (1978b). Mechanisms of developmental integration of *Aster novae-angliae* and *Hieracium murorum, Ann. Bot.* **42**, 561–575.

Frijters, D., and Lindenmayer, A. (1976). Developmental description of branching patterns with paracladial relationships, in *Automata, Languages, Development* (edited by A. Lindenmayer and G. Rozenberg), North-Holland, Amsterdam, pp. 57–73.

Konrad Hawkins, E. (1964a). Developmental studies in regenerates of *Callithamnion roseum* Harvey. I The development of a typical regenerate, *Protoplasma* **58**, 42–59.

Konrad Hawkins, E. (1964b). Developmental studies in regenerates of *Callithamnion roseum* Harvey. II Analysis of apical growth. Regulation of growth and form, *Protoplasma* **58**, 60–74.

Lindenmayer, A. (1968). Mathematical models for cellular interactions and development. II Simple and branching filaments with two-sided inputs, *J. theor. Biol.* **18**, 300–315.

Lindenmayer, A. (1971). Developmental systems without cellular interactions, their languages and grammars, *J. theor. Biol.* **30**, 455–484.

Lück, H. B. (1975). Elementary behavioural rules as a foundation of morphogenesis, *J. theor. Biol.* **54**, 23–34.

Lück, J., and Lück, H. B. (1974). Production raméale et forme, *Rev. Cytol. Biol. Veg.* **37**, 313–322.

Lück, H. B., and Lück, J. (1976). Cell number and cell size in filamentous organisms in relation to ancestrally and positionally dependent generation times, in *Automata, Languages, Development* (edited by A. Lindenmayer and G. Rozenberg), North-Holland, Amsterdam, pp. 109–124.

Chapter 20

Simulation of growth with anastomosis: a colonial hydroid

The frontispiece of this book shows a nine-day-old colony of the polymorphic marine hydroid *Podocoryne carnea.* This hydroid usually grows on snail shells inhabited by hermit crabs, but this example has been grown on a microscope slide, to give a plane two-dimensional pattern. A single hydranth from a colony is set down on the slide in standing sea water, becomes attached within a few days and sends out stolons, which branch and anastomose. New hydranths appear on these stolons and feed on detritus. Growth involves cell migration from older parts of the colony, in which cells divide, to regions of active growth such as the tips of stolons. A paper by Braverman (1974) reviews an extended investigation of this organism. As part of this investigation the growth was simulated, on a rectangular grid, by Braverman and Schrandt (1967).

The colony is by no means like a tree; there is extensive anastomosis. In six photographs of nine-day colonies of similar extent to that illustrated, on the average 57 out of 125 edges belong to closed loops. (These photographs, and others referred to in this chapter, were kindly provided by Dr. M. H. Braverman.) It is known that when a growing stolon tip approaches another stolon lying across its path it is not merely stopped, but fuses with the other so that transfer of cells can take place between them. Indeed it appears the approaching stolon tip causes a branch point to form on the side of the impeding stolon, and the two tips fuse (Braverman 1974).

20.1 CONSEQUENCES OF ANASTOMOSIS

Anastomosis has several consequences for modelling. Perhaps the most important is that it becomes harder to test local hypotheses about growth in terms of single pictures of colonies of specified age. In the photograph of this colony it is hard to tell a bifurcation from a point of anastomosis in a closed loop of stolon segments. So it becomes more important to have a sequence of dated photographs of each colony studied. L-system theory, as discussed in the previous two chapters, becomes irrelevant. Some of its generalizations, such as those discussed by certain contributors to Lindenmayer and Rozenberg (1976), are potentially applicable to network modelling. However, at present this work is primarily mathematical and will not be reviewed here. The ordering methods on which much of the work reviewed in this book depends are largely irrelevant here. One can investigate the non-anastomosing fringes of colonies, counting terminal units in the manner of Hollingworth and Berry (1975) which is discussed in Section 17.1. Looking at nine examples of nine-day colonies and ten of older colonies, I find in all only twenty-two units with four branches of Strahler order 1, and seventeen units with five branches of Strahler order 1, which are free from anastomosis. This is not enough to obtain a significant comparison with segmental and terminal growth models. Examination of the photographs shows that short side branches are not uncommon within closed stolon loops, indicating segmental growth. The central region of the colony gets rather dense as the colony grows, indicating that a rule of loop area conservation, as employed for example by Ortmann and Harte (1976) for patterns of leaf veins, is not appropriate.

Features partially analogous to anastomosis can be encountered even in simulating structures which are free from closed loops. It may be appropriate to assume that the growth of a branch tip is inhibited by proximity to other branches or other branch tips. Honda *et al.* (1981) incorporate this kind of inhibition in a model of horizontal tiers of bifurcating branches in species of *terminalia*. At each step in the simulation a horizontal circle is drawn around each terminal point P. If another such point P′ lies within this circle the point P does not branch. This greatly reduces tangling of branches.

20.2 SIMULATIONS OF *PODOCORYNE CARNEA*

Turning to the actual simulations of Braverman and Schrandt (1967), it is important to keep in mind that one wishes if possible to test local hypotheses about growth by the comparison of global features of real and simulated colonies. The simulation is carried out on a rectangular grid. At each time step one first applies a probabilistic growth rule to each stolon tip, and accordingly moves the tip forward by one grid step or leaves it unextended. One then applies probabilistic rules for branching to each stolon segment, at each internal grid point on the segment either doing nothing, adding the first grid step of a new branch to one side, or adding such a step to the other side. All moves made in these first two stages are next checked for the crossing of one stolon by another, and the growth step cancelled if necessary. Finally a probabilistic rule is applied for the appearance of new hydranths.

The rules adopted for stolon growth are based on the concept that proximity to hydranths promotes growth. Let d_{ij} denote the minimum distance from the ith hydranth to the jth stolon tip. (Because there are closed loops the distance from a hydranth to a stolon tip can be ambiguous.) One rule used is that the growth rate of the jth tip is proportional to the quantity

$$\sum_i (d_{ij} + 1)^{-1}$$

Another is that the growth rate of the jth tip is proportional to the number of d_{ij} values less than some fixed value. A branch point is assumed to arise, at a random location on the network, whenever a fixed number of new steps have been made by stolon tips since the last branch point appeared. This can be interpreted as a consequence of a new branch point giving a single burst of some agent which inhibits further branching, with the concentration of the inhibitor falling as the network grows. New hydranths are assumed to be inhibited in the vicinity of all existing hydranths.

As a detailed example of the simulations, one set of patterns were generated using the following set of rules:

1. two-dimensional growth on a mesh of unit squares;
2. stolons can grow N, E, S or W from the seeding hydranth;
3. unit step growth in each time interval;
4. all stolons are straight;
5. a tip grows a new unit if there are five or more hydranths within 5 units from the tip. If there are fewer, the probability of a new step is their number, divided by five;
6. a new hydranth is inserted when there are three adjacent grid points on the network with no hydranth;
7. for every four new steps added, and for every anastomosis, a new branch point is added.

Braverman and Schrandt (1967) illustrate an example of a colony produced in 34 generations applying these rules. Out of 470 segments, only 71 do not belong to loops. They also illustrate an example of a colony produced in 34 generations using a modified rule 7. A new branch point is not provided when anastomosis occurs. This gives a more open pattern (shown in Fig. 20.1) but again only 47 segments out of 263 do not belong to loops. In these examples the proportion of segments in loops is much higher than in the nine-day colonies (***Frontispiece***). Comparison should, however, be made with older, larger colonies. Unfortunately in the available photographs of older colonies much of the network is obscured by the presence of large sexual zooids.

20.3 SOME POSSIBLE EXTENSIONS

Braverman and Schrandt (1967) compare simulated and real colonies only in terms of the total numbers of occupied mesh points and of hydranths. They do not compare the geometry of the real and simulated colonies. Also the later work of Braverman (1971, 1974) shows up some points in which the assumptions of these simulations are incorrect. He was able to grow a colony of which half was fed and half starved. Both halves grew at equal pace, indicating that it is not appropriate

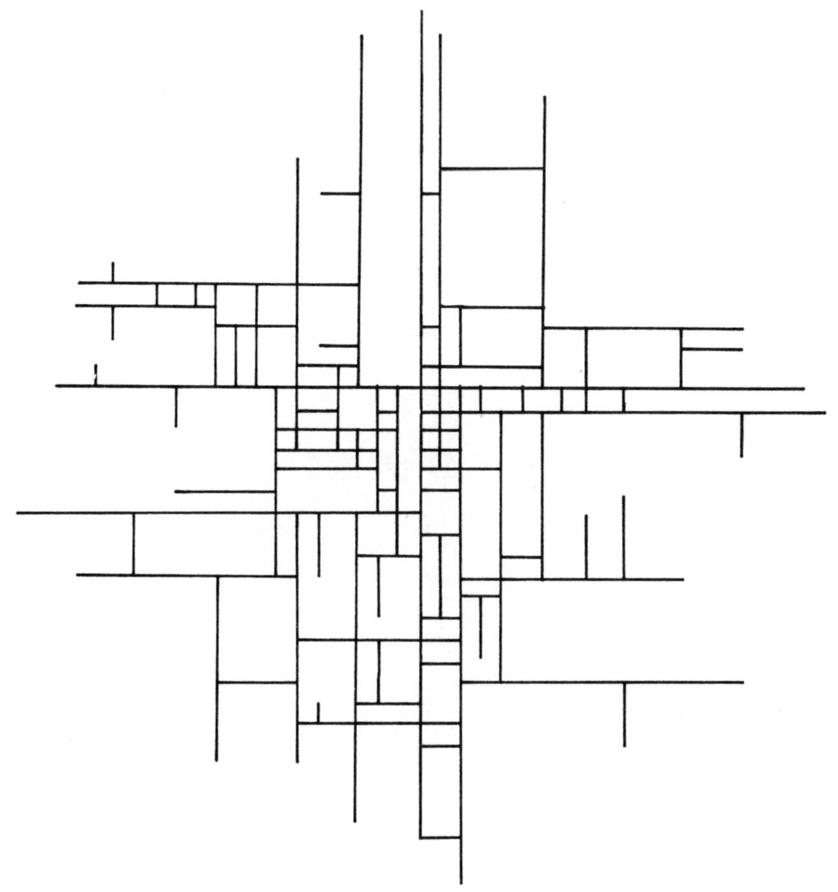

Fig. 20.1 A simulated colony of *Podocoryne carnea*. Reproduced, with the symbols for hydranths omitted, from 'Colony development of a polymorphic hydroid as a problem in pattern formation', *General Systems*, **12**, 39–51, by M. H. Braverman and R. G. Schmidt, by permission of the Society for General System Research. © 1967 by the Society for General System Research

to take local growth depending on proximity of hydranths. He also found that most branching occurs on stolons which are no more than two days old.

It thus remains an open topic how far successful simulations can be made of these colonies, with growth rules not using proximity to hydranths and with stolon aging affecting branching probability. Also the possible usefulness of measures of the statistical geometry of the real and simulated colonies needs to be investigated. Such measures have been introduced in geography (Haggett and Chorley 1969) and in metallurgy and mineralogy (Serra 1982). Typically the use of these measures is based on treating the network as a sample of an infinite network, and modifications may be needed to deal with the finite extent of the colonies and their simulated versions.

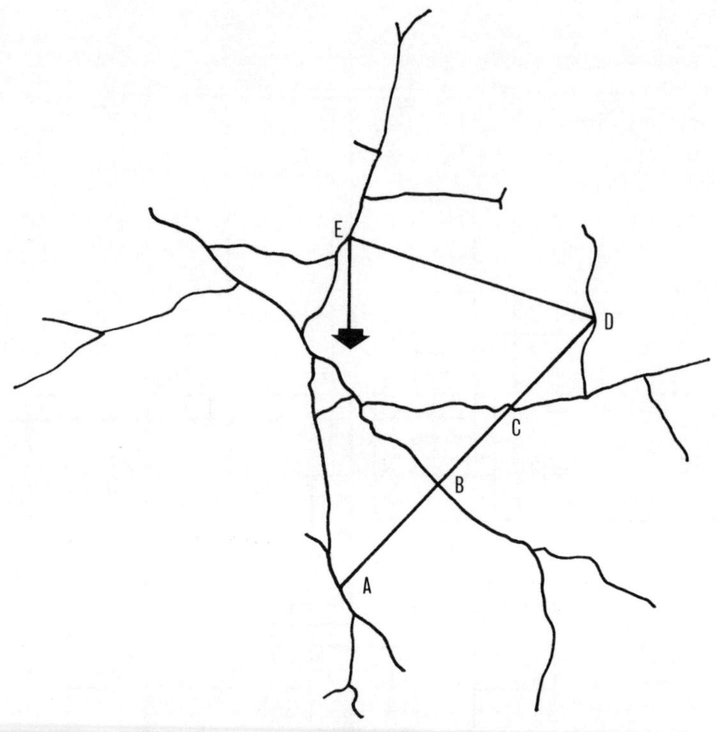

Fig. 20.2 The transect method of Dacey. The transect starts at an arbitrary point A on a stolon, and proceeds in an arbitrary direction to cut stolons at B, C and to meet another at D. If extended further, it would not encounter another stolon. So a new arbitrary direction is selected on which the transect encounters a stolon at E, and so on

Point patterns have been more extensively studied than network patterns, and some aspects of the colonies, such as the distribution of hydranths, are point patterns. Indeed the stolon network itself can be seen to give rise to a point pattern, employing a transect method due to Dacey (1967). This is illustrated in Fig. 20.2, for an arbitrary simple network. One first picks a point A on the network at random, and draws a transect in a random direction, extending it to the point D beyond which it leaves the network. Then from D one takes a second transect in a random direction, following it to E, and so on. The distances AB, BC, CD, DE, EF, . . . are recorded and treated as distances between a sequence of points on a line. The nearest neighbour statistics of random points on an infinite line follow a simple rule (Clark 1956). A pair of points X,Y are reflexive nearest neighbours if X is the nearest point to Y and Y is the nearest point to X. They are reflexive nth neighbours if X is the nth nearest point to Y and Y is the nth nearest neighbour to X. For random points on an infinite line, the proportion of reflexive nth nearest neighbours falls as

$$(2/3)^n$$

Sets of points tending to regular spacing give values exceeding this, while sets tending to cluster give smaller values. In the colony patterns end effects are likely to be important, and one should compare not with this simple result but with the results obtained from transects on arbitrary patterns of finite extent, such as a pattern of radial lines and a regular mesh pattern.

I have already used a measure of the importance of anastomosis in the topology of these colonies, namely E_1/E, where E is the total number of edges and E_1 is the number of edges that belong to closed loops. A geometrical analogue of this would be the ratio A_1/A, where A is the total area of the colony, defined by drawing straight lines between the tips of external branches, and A_1 is the area of closed loops. It might be found advisable to construct, for the purposes of comparisons of the statistical geometry of simulations and real colonies, arbitrary finite patterns with the observed values of E_1/E or of A_1/A. (Scrabble games yield rather small—E about 50—examples with E_1/E comparable to the values in the 9-day colonies.) Any statistical index not distinguishable between a colony simulation and such a 'Scrabble' pattern would be of little value in assessing the merits of the simulation.

References

Braverman, M. H. (1971). Studies in hydroid differentiation. IV Cell movements in *Podocoryne carnea, J. exp. Zool.* **176**, 361–382.

Braverman, M. H. (1974). The cellular basis for colony form in *Podocoryne carnea, Amer. Zool.* **14**, 673–698.

Braverman, M. H., and Schrandt, R. G. (1967). Colony development of a colonial hydroid as a problem in pattern formation, *General Systems* **12**, 39–51.

Clark, P. J. (1956). Grouping in spatial distributions, *Science* **123**, 373–374.

Dacey, M. F. (1967). Descriptions of line patterns, *Northwestern Studies in Geography* **13**, 277–287.

Haggett, P., and Chorley, R. J. (1969). *Network Models in Geography*, Methuen, London.

Hollingworth, T., and Berry, M. (1975). Network analysis of dendritic fields of pyramidal cells in the neocortex and Purkinje cells in the cerebellum of the rat, *Phil. Trans. Roy. Soc.* **B270**, 227–264.

Honda, H., Tomlinson, P. B., and Fisher, J. B. (1981). Computer simulation of branch interaction and regulation by unequal flow rates in botanical trees, *Amer. J. Bot.* **68**, 569–585.

Lindenmayer, A., and Rozenberg, G. (1976). *Automata, Languages, Development*, North-Holland, Amsterdam.

Ortmann, A., and Harte, C. (1976). Development of the pattern of veins in the leaf of *Antirrhinum majus* L., in *Automata, Languages, Development* (edited by A. Lindenmayer and G. Rozenberg), North-Holland, Amsterdam, pp. 89–95.

Serra, J. (1982). *Image Analysis and Mathematical Morphology*, Academic Press, New York.

Appendix 1

Simplicial complexes

To change from a description of a foodweb in terms of the resource graph to a description in terms of a simplicial complex, as discussed in Section 4.4, we have to interpret the dominant cliques as polyhedra in an abstract space. In the resource graph a dominant clique is a set of n points with all the $\frac{1}{2}n(n-1)$ pairs joined by lines. This can be represented in a two-dimensional picture such as Fig. 4.3 because the lines are allowed to cross. In the alternative representation we have again a set of n points, but these are interpreted as the vertices of a polyhedron in a space of dimension at least $n-1$. Thus a clique of one member is a point, while one with two members is a line, and one of three members is a triangle. A clique of four members is represented not by a square with both its diagonals, but by a tetrahedron, requiring a space of three or more dimensions. These polyhedra are called simplexes, and a collection of them form a simplicial complex. This is a solid object made by gluing together the individual simplexes at a common point, along a common edge, and so on, according as the dominant cliques share one, two, or more members. If two distinct dominant cliques share four members, then each of them must possess at least five members. So each must be a polyhedron in a space of at least four dimensions, and one can glue them together on a three-dimensional common hyperface.

It makes intuitive sense to say that such a complex can have one or more holes in it. In Fig. A1.1 there are three triangular simplexes, A, B, and C. In Fig. A1.1**A** there is a hole—the central triangle—while in **B** there is not. In a foodweb for which the resource graph contains cliques corresponding to A, B, and C, the component A eats components 1, 2, and 3, the component B eats 3, 4, and 5 and the component C eats 4, 6, and 2, in the example with a hole. In the example with no hole, A still eats 1, 2, and 3, and B still eats 3, 4, and 5, but C eats 2, 3, and 4. The component C has been added conservatively in **B** but non-conservatively in **A**, in the sense defined in Section 4.1.

For a complex built up of many simplexes which may be of high dimension, there are algebraic methods to determine the presence or absence of a hole.

Two simplexes of dimension n can be connected in $n-1$ different ways, according as the common face has dimension $n-1$, $n-2$, ... down to a line or a point.

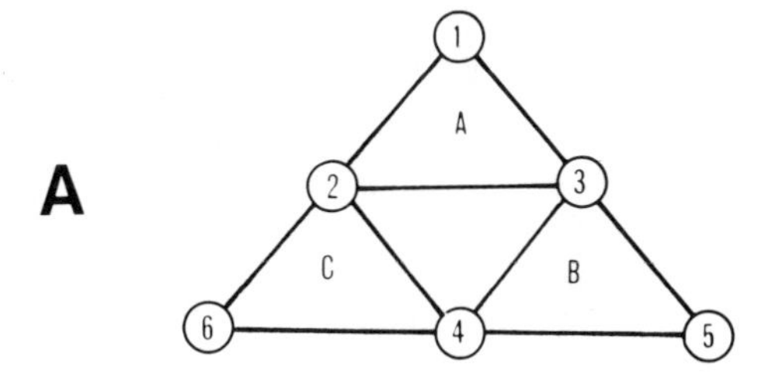

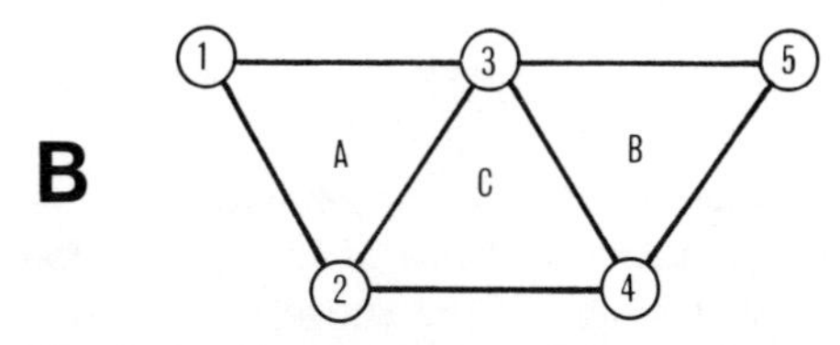

Fig. A1.1 Simplicial complexes built up from triangular simplexes, A, B and C. In **A** there is a hole, in **B** there is not

Assigning a measure of complexity to a simplicial complex will therefore require consideration of the dimensions of the simplexes, the number of connections, and the dimensions of these connections. Casti, in the book cited in Chapter 5, has set down particular complexity measures and evaluated them for a single artificial foodweb, drawn from an elementary zoology textbook. These measures are complicated and it is not clear that they have any real advantages over simpler ones appropriate if the foodweb is treated as a digraph.

Appendix 2

Linear homogeneous difference equations

In this book linear homogeneous difference equations arise in two quite distinct contexts. In Chapter 7 I examined the local equilibrium point x^0 of a non-linear difference equation. For example, if the equation is of second order

$$x(t + 2) = F(x(t), x(t + 1))$$

x^0 satisfies

$$x^0 = F(x^0, x^0)$$

and near $x = x^0$ the difference, $X = x - x^0$, satisfies approximately the linear homogeneous equation

$$X(t + 2) + a\,X(t + 1) + b\,X(t) = 0 \tag{A2.1}$$

In Section 18.3 such a second-order equation arises exactly from the recurrence relation for the lengths of successive strings generated by an L-system. In example 1 in Section 18.3, the difference equation is equation (18.9), which is equation (A2.1) with $X = f_G$, $a = -1$, and $b = -1$. In example 2 defined by expression (18.1), the difference equation is equation (18.12), which is equation (A2.1) with $X = f_{\mathrm{A,B}}$, $a = -1$, and $b = -2$. Alternatively this example gives the inhomogeneous equation (18.13) for $f_{(,)}$, but this can be trivially reduced to the same example of equation (A2.1) with X now taken to be $f_{(,)} + 1$.

The general solution of equation (A2.1) is the sum of two exponentially growing, or decaying, terms in proportions fixed by the values $X(1)$ and $X(2)$. Thus

$$X(t) = A_1 \lambda_1^t + A_2 \lambda_2^t \tag{A2.2}$$

where λ_1 and λ_2 are the roots of the secular equation

$$\lambda^2 + a\lambda + b = 0 \tag{A2.3}$$

These are

$$\lambda_{1,2} = \frac{-a}{2} \pm \tfrac{1}{2}[a^2 - 4b]^{\frac{1}{2}} \tag{A2.4}$$

For local stability we require that both λ_1 and λ_2 have modules less than 1, since otherwise one term (at least) in the expression (A2.2) grows continually.

To solve the examples in Section 18.3, we use (A2.2) and (A2.3) together with

$$\begin{aligned} A_1\lambda_1 + A_2\lambda_2 &= X(1) \\ A_1\lambda_1^2 + A_2\lambda_2^2 &= X(2) \end{aligned} \tag{A2.5}$$

We obtain for the Fibonacci sequence of example 1

$$f_G(t) = 5^{-\frac{1}{2}}2^{1-t}[(1 + 5^{\frac{1}{2}})^{t+1} - (1 - 5^{\frac{1}{2}})^{t+1}]$$

and for the two growth functions of example 2

$$f_{\mathrm{A,B}}(t) = 2^t$$

$$f_{(,)}(t) = \tfrac{1}{3}[2^t - (-1)^t] - 1$$

Appendix 3

Laminar flow in a tube

The general equation for the laminar flow of a viscous incompressible fluid in the absence of gravity is

$$\boldsymbol{\nabla} p = \mu \boldsymbol{\nabla}^2 \boldsymbol{v}$$

where p is pressure, μ is viscosity and $\boldsymbol{v}$ is velocity. For flow along a uniform cylindrical tube oriented along the $\hat{\boldsymbol{z}}$ direction

$$\boldsymbol{v} = V(r)\,\hat{\boldsymbol{z}}$$

and the equation is

$$P = \mu\left(\frac{d^2 V}{dr^2} + \frac{1}{r}\frac{dV}{dr}\right)$$

where P is the pressure gradient. $V(r)$ is assumed to be finite along the axis ($r = 0$) and zero at the wall ($r = d/2$) so that the appropriate solution of this last equation is

$$V(r) = -\frac{P}{4\mu}(d^2/2 - r^2)$$

The flow, which is the volume crossing any normal cross-section in unit time, is

$$Q = 2h\int_0^{d/2} rV(r)dr = \frac{\pi d^4}{128\mu}P = \frac{\pi d^4 \Delta P}{128\mu l}$$

where ΔP is the pressure drop along a tube of length l. Resistance is defined as the ratio of pressure drop to flow

$$\frac{128\mu l}{\pi d^4}$$

The shear stress exerted by the fluid on the wall surface is equal to the pressure gradient times the radius, and the shear for a unit length of tube is the circumference times this shear stress, which comes to

$$\frac{32\mu Q}{d^2}$$

Appendix 4

Details of the analysis of branching angles

In order to analyse branching angles as discussed in Section 15.2, we wish to maximize a cost function of the form

$$F = \sum_i l_i f(d_i, q_i) \tag{A4.1}$$

where the index i takes the values 0, 1, 2 referring as in Fig. 13.1 to the three vessels meeting at the bifurcation. The diameters d_i and the flows q_i are regarded as fixed. The lengths l_i are the distances along the central axes of the vessels from the bifurcation point (X, Y) to the end points (X_0, Y_0), $(X_1 Y_1)$ and (X_2, Y_2), as shown in Fig. A4.1. The end points are kept fixed and the position of the point of bifurcation is varied. So we look for a minimum of F with respect to variation of X

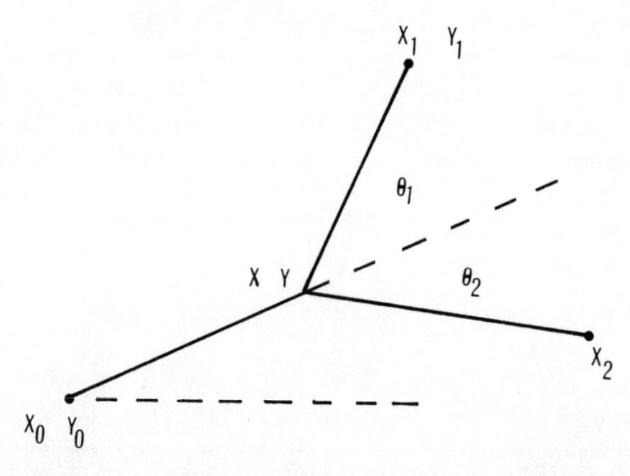

Fig. A4.1 Geometry of the central axes of a parent vessel and two daughters, meeting at a bifurcation point

and Y. For a turning value of F we require the equalities

$$\frac{\partial F}{\partial X}=0, \frac{\partial F}{\partial Y}=0 \tag{A4.2}$$

For a minimum we also require two inequalities

$$\frac{\partial^2 F}{\partial X^2}>0$$

$$\frac{\partial^2 F}{\partial X^2}\frac{\partial^2 F}{\partial Y^2}>\frac{\partial^2 F}{\partial X \partial Y} \tag{A4.3}$$

From Fig. A4.1 we see that

$$\begin{aligned} l_0 &= [(X - X_0)^2 + (Y - Y_0)^2]^{\frac{1}{2}} \\ \cos\theta_0 &= (X-X_0)/l_0 \\ \sin\theta_0 &= (Y-Y_0)/l_0 \end{aligned} \tag{A4.4}$$

with similar results for l_1, l_2, and the angles $(\theta_1 + \theta_0)$ and $(\theta_2 - \theta_0)$. From equations (A4.4) we can evaluate the derivatives

$$\frac{\partial F}{\partial X}=f_0 \cos\theta_0 - f_1 \cos(\theta_1 + \theta_0) - f_2 \cos(\theta_2 - \theta_0) \tag{A4.5}$$

and

$$\frac{\partial F}{\partial y}=f_0 \sin\theta_0 - f_1 \sin(\theta_1 + \theta_0) + f_2 \sin(\theta_2 - \theta_0) \tag{A4.6}$$

so that we can apply equations (A4.2). These fix the angle $(\theta_2 - \theta_0)$ in terms of the angle $(\theta_1 + \theta_0)$, and give $\cos(\theta_1 + \theta_0)$ as the solution of a quadratic equation in terms of the tree cost functions f_0, f_1, and f_2 and the angle θ_0. However, the results cannot have a significant dependence on the arbitrary choice of a direction from which θ_0 is measured, so one can set this angle equal to zero. Then $\cos\theta_1$ is given as the solution of a quadratic and so can be expressed as an explicit function of the cost functions. One then has to verify that one of the two solutions of the quadratic is consistent with the inequalities of equations (A4.3).

The outcome of this laborious process is the pair of equations

$$\begin{aligned} \cos\theta_1 &= \frac{f_0^2 + f_1^2 - f_2^2}{2 f_0 f_1} \\ \cos\theta_2 &= \frac{f_0^2 + f_2^2 - f_1^2}{2 f_0 f_2} \end{aligned} \tag{A4.7}$$

and finally the principal result is for the total angle of bifurcation

$$\cos(\theta_1 + \theta_2) = \frac{f_0^2 - f_1^2 - f_2^2}{2f_1f_2} \tag{A4.8}$$

Appendix 5

Impedance and the reflection of waves

To understand the nature of the calculations in the models of Taylor and of Sidell and Fredberg, discussed in Section 16.3, it is necessary to appreciate the role of the impedance in wave reflection. I shall describe a simple case, that of sound waves of wavelength λ in tubes with rigid walls and constant cross-sectional areas S_i, where $S_i \ll \lambda^2$. In terms of the molecular displacement $\xi(x, t)$, the wave equation is

$$\frac{\partial^2 \xi}{\partial t^2} = \frac{1}{c^2} \frac{\partial^2 \xi}{\partial x^2} \tag{A5.1}$$

where the velocity c satisfies $c^2 = dP/d\rho$, P being the local pressure and ρ the local density. In terms of the mean pressure P_0 and the mean density ρ_0, the force exerted by the fluid as a result of the presence of the sound wave is

$$S_i(P - P_0) = S_i c^2(\rho - \rho_0) = -S_i c^2 \rho_0 \frac{\partial \xi}{\partial x} \tag{A5.2}$$

Now consider a tube extending from $x = -L$ to $x = 0$, with a load at $x = 0$. This load could be a piston, or it could be another tube, or tubes, extending from $x = 0$. The impedance Z of the load is defined by

$$Z \left.\frac{\partial \xi}{\partial t}\right|_{x=0} = -S_0 \rho_0 c^2 \left.\frac{\partial \xi}{\partial x}\right|_{x=0} \tag{A5.3}$$

It is convenient also to define a dimensionless quantity, the relative impedance $\tilde{Z}$, as $\tilde{Z} = Z/S_0 \rho_0 c$, so that

$$\tilde{Z} \frac{\partial \xi}{\partial t} = -c \frac{\partial \xi}{\partial x} \tag{A5.3'}$$

For an incident harmonic wave $\exp(i\omega(t - x/c))$, and a reflected wave

$$\beta \exp(i\omega(t + x/c) + i\phi) = B \exp(i\omega(t + x/c))$$

the response of the load can be characterized, using equation (A5.3′), by

$$B = \frac{\tilde{Z} - 1}{\tilde{Z} + 1}, \tilde{Z} = \frac{1 - B}{1 + B} \tag{A5.4}$$

For example, if the load is a piston of mass M held by a spring with spring constant k and subject to linear viscous damping of strength b, the application of Newton's third law yields

$$Z = b + i\left(M\omega - \frac{k}{\omega}\right)$$

If there is no damping, so that no energy is lost, Z is purely imaginary. Then $\beta = 1$ and the phase ϕ of the reflected wave depends on M and k in a manner sensitive to the frequency ω. The mass resonates when ω coincides with the natural frequency, that is when $M\omega = k/\omega$. Resonance in general is characterized by a sign change in the imaginary part of the impedance and a minimum in the magnitude of the impedance.

If the load is an infinite extension of the original tube, with the same cross-section S_0, there can be no reflected wave and $\tilde{Z} = 1$. So we can define the impedance of an infinite tube of cross-section S_0 as $Z = S_0 \rho_0 c$. If the load is an infinite tube of a different cross-section S_1, then as 'seen' from the first tube this has relative impedance $\tilde{Z} = S_1/S_0$. This is a real quantity; some of the energy is transmitted down the second tube so that $\beta < 1$.

If the load is a rigid wall, so that there is no displacement at $x = 0$ and consequently $B = -1$, the impedance is infinite.

If we know the relative impedance $\tilde{Z}$ of a load at the right-hand end $x = 0$, and we also know the length of the tube L and the sound velocity c, we can evaluate the corresponding quantity at $x = -L$, by using equation (A5.4) and the incident and reflected amplitudes at this end. So we can define the 'left-hand' impedance of the tube and its load, Z_-. We can now treat a loaded tube as the load at the right-hand end of another tube, and go on to deal with a sequence of connected tubes of various cross-sections S_i. They need not be aligned; this is a consequence of the approximation of long wavelength. To determine the final amplitude of the reflected wave at the left-hand end of the sequence of tubes, we need the values of S_i and of the lengths L_i, and the load on the final (right-hand) tube.

The final step needed to follow the kind of calculation carried out by Taylor is to find the effect of a bifurcation. This is passed over very swiftly in his account. Again because of the long wavelength approximation, the angles θ_1 and θ_2 are immaterial and only the impedances of the outgoing branches 1 and 2 matter. But how are they to be combined? The pressures in all three tubes are the same, so that

$$\left.\frac{\partial\xi}{\partial x}\right|_0 = \left.\frac{\partial\xi}{\partial x}\right|_1 = \left.\frac{\partial\xi}{\partial x}\right|_2 \tag{A5.5}$$

No fluid is lost, so

$$S_0 \left.\frac{\partial\xi}{\partial t}\right|_0 = S_1 \left.\frac{\partial\xi}{\partial t}\right|_1 + S_2 \left.\frac{\partial\xi}{\partial t}\right|_2 \tag{A5.6}$$

Equations (A5.5) and (A5.6) imply that the relative impedance of tubes 1 and 2 as a load at the left-hand end of tube 0 is $\tilde{Z}_0$ satisfying

$$\frac{S_0}{\tilde{Z}_0} = \frac{S_1}{\tilde{Z}_{1-}} + \frac{S_2}{\tilde{Z}_{2-}} \tag{A5.7}$$

By analogy with the theory of electrical circuits, we can say that the tubes 1 and 2 act in parallel. If the tubes 1 and 2 are infinite, and so have $\tilde{Z}_{1-} = \tilde{Z}_{2-} = 1$, then the condition for no reflection, $\tilde{Z}_0 = 1$, is that the total cross-section is constant

$$S_0 = S_1 + S_2$$

Taylor's calculation proceeds in this way. He first calculates the impedances of loads at the ends of branches of Strahler order 1. Then he calculates the 'left-hand' impedances of these branches. He uses a symmetrical model, so these branches all combine in pairs, and for each pair he uses the parallel addition result (A5.7), to get the impedances of loads on the branches of Strahler order 2. He continues until the whole symmetrical tree is assembled and thus evaluates the impedance of the tree. Sidell and Fredberg carry out a similar process for an asymmetrical tree using the regular model of Horsfield, in which the ratios S_0/S_1 and S_0/S_2 can be taken as the same at each bifurcation once branches of Horsfield order $\Delta + 1$ are reached.

If there is no damping, as in Taylor's calculations with tubes having perfectly elastic walls, the effect of this elaborate process is to make the final impedance a complicated function of the frequency. An incident pulse, which is a superposition of harmonic waves with a continuous range of frequencies, gives rise to reflected waves with phases depending in a complicated manner on their frequency. These combine to form an outgoing signal very different from the incident pulse. There is disguised reflection rather than true absorption.

Subject Index

Where a page number is printed **bold**, this refers to the definition of a term in graph theory.

Where a keyword is a major topic of a chapter or section, only the first page of that chapter or section is given.

Binomial species names that only occur in tables are not indexed.

adjacency matrix, **4**
adjacent, **4**
airways, 106, 130, 140, 152, 166
  branching angles, 143, 155, 157
  data in Strahler ordering, 129, 130
  diameters, 151
algae, branching, 185
  filamentous, 184
alphabet, 178
alveoli, 106, 150, 166
amensalism, 57
anastomosis, 2, 109, 131, 194
ants, dominance in, 11
apex, vegetative, 190
arc, **4**
arterial branching angles, 155, 206
  data in Strahler ordering, 128, 130
  diameters, 136, 151
arteries, 108, 130, 131, 137, 151, 155, 157, 165
  in retina, 137, 158
Aster novae-angliae, 189, 192
asymmetrical tree, 107, 146, 152, 158, 211

bacteriophage, 98
binary tree, **5**
Boolean analogue functions, 81
  function, 81
    forcing, 95
  rate equations, 81, 89
  variables, 81
botanical tree data, 122, 124, 125, 131, 133, 150, 160
branch, **111**
branch weight, 131
branching angles, arterial, 155, 206
  in airways, 143, 155, 157
bronchi, 106
bronchial data in Strahler ordering, 129, 130
bronchioles, 106
bundles, 160

cable equation, 163
Callithamnion roseum, 185
casts, arterial and bronchial, 137
centrifugal ordering, 103, 112, 173
centripetal ordering, 103, 112
Chaetomorpha linum, 185
clique, **35**
  dominant, **35**
Codium fragile, 189
colonial animals, 130, 194
community web, 21, 37
compartmentalization, 53
competition, 11, 31, 67
  by sharing prey, 31
  compared with predation, 67
complexity, 40
  and information theory, 44
  and stability, 49
  NC as measure of, 52

of algorithm, 40
of food web, 20, 52, 201
of graph, **42**
of snowflake, 40
of tree, 42, 119
consecutive ones property, **35**
conservative addition, 31, 200
consumer graph, **32**
consumer interval foodweb, 37
context-free L-system, 178, 182, 185, 189, 190
rule, 177
context-sensitive L-system, 178, 186, 193
rule, 177
contiguous, **5**
control processes, biochemical, 52, 72, 79
genetic, 89, 98
coral reef species, 11
cost function, 151, 155, 206
cross-section, total, 104, 107, 108, 154, 161, 164, 166
cumulative ordering, 112
cycle, **5**
cycle analysis of stability, 50, 66
cycles, disjoint, 58
in clocked switching networks, 89

DOL system, 178, 182, 185
delays, 69
dendritic data in Strahler ordering, 128, 136
diameters, 164
trees, 104, 130, 137, 163, 168
difference equations, 56, 62, 202
differential equations, ordinary, 3, 50, 56, 78
digraph, **4**
dynamic, 17, 82
static, 12
dimension, 140
Dirichlet polygon, 150
divaricating shrubs, 125, 126, 130
dominance, 2, 11
patterns, 11
reversal, 16
donor control, 50, 75
dynamic component digraph, **82**
dynamic interaction digraph, **17**

E. coli, 98
edge, **4**
exterior, **42**
elastic similarity, 163

entropy, 152
excess branches, 116

Fahraeus–Lindquist effect, 154
feasible equilibrium point, 52
Fibonacci sequence, 181, 203
fluency, 93
food chain, 23
canonical, 25
redundancy, 44
food web, 2, 20, 44, 49
community, 21, 37
consumer interval, 37
multihabitat, 20
North Sea, 25
resource interval, 37
sink, 21
source, 21
strictly trophic, 21
weighted, 23
forcing, 94
Boolean function, 95
cycle, 97
digraph, **96**
structure, 96
fossil plant data, 127, 130, 131, 138, 163
freshwater communities, 21, 28, 32, 50
fungi, 130, 173

Gause exclusion principle, 67
geography, methods from, 110, 197
Goodwin oscillator, 72, 83
graph, **4**
Gravelius ordering, 110
growth cone, 169
growth function, 181

Hieracium murorum, 189, 192
hierarchy, 11, 54
index, 14
Hirudo medicinalis, 85
horn, 166
Horsfield ordering, 113, 146, 166
Horton ordering, 111
law of branching ratio, 115
Horton's laws, 110, 119
hydroid, 194
hypercube, 17, 82

impedance, 165, 209
inflorescence, 189
inhibition of branch tips, 195
function, 78

internal homogeneity, 93
interval analysis of food webs, 31
 of gene, 37
interval graph, **35**
interval property, 31
intrinsic traits and dominance, 15

Jacobian matrix, 51, 56, 70

Kauffman networks, 89, 94

L-systems, 175, 184
 deterministic, 178
 (m, n), 178
 table, 178, 190
laminar flow, 152, 155, 204
leaf, **5**
length of cycle, **5**
Lindenmayer systems, 175, 184
liveliness, 95
loop, **5**
Lotka–Volterra models, 51, 79

macacque monkeys, dominance in, 11, 15
Markov process, 16
matrix, adjacency, **4**
 Jacobian, 51, 56, 70
 random, 52
memory, 93
motoneurons, 164

neighbour, reflective, 198
neuron firing, 85
niche overlap, 31
node, **4**
North Sea food web, 25

omnivory, 53
ordering schemes, centrifugal, 103, 112, 173
 centripetal, 103, 112
 cumulative, 112
 Gravelius, 110
 Horsfield, 113, 146, 166
 Horton, 111
 Shreve, 111, 186
 Strahler, 109, 110, 118, 122, 158, 168
 Weibel, 107, 113, 187
orders of branching, 2, 103
 number in lung, 107, 115, 166
out-degree, **13**

paracladial tree, 175, 191
path, **5**
pecking order, 2, 11
Picea Glehni, 160
pipe model, 131, 160
plant–aphid–parasitoid communities, 52
Podocoryne carnea, 131, 194
power laws in biochemistry, 79
 in botany, 131, 149, 160
 in o.d.e. models, 78
 in physiology, 107, 131, 140, 151, 164
 in volume filling trees, 140
predation, 2, 20
 compared with competition, 67
1 predator + 2 prey, 70
production rules, 178
Purkinje cells, 130, 168
pyramidal cells, 104, 168

Rall model, 163
random graph in sense of Erdos, 98
 of Shreve, 117
 matrix, 52
 trees in growth models, 171
 trees in plane, 117
 walk, 143
rank, 11
rate equations, 50, 78, 80
recurrence in Shreve orders, 186
 relations, 180, 185, 189
 system, 180
reflection, 165, 209
regular model, 148, 166, 211
replication time, 89
resource, continuous, 31
 discrete, 31
 graph, **35**
 interval food web, 37
respiration, 24
rigid circuit, **35**
root, **5**
rooted tree, **5**
 binary tree, 103
rosette, **59**, 62, 69, 74
Routh–Hurwitz determinant, 61
Roux' rules, 155

score sequence vector, 13
secular equation, 51, 58, 60, 73
 mnemonic, 60
seedbank, 74
segment, 130
segmental growth model, 171, 195
self-similarity, 119
self-stabilization, 53, 74
semicircle law, 52

shear, 155, 205
Shreve ordering, 111, 186
signed digraph, **5**
similarity, elastic, 163
  self, 119
  stress, 163
similarity dimension, 140
simplicial complex, 39, 200
simulation of algae, 184
  anastomosing network, 194
  botanical trees, 150, 168, 195
  dendrites, 168
  Podocoryne carnea, 194
  using L-systems, 175, 184
sink web, 21, 37
source web, 21, 37
spanning tree, **43**
stability and complexity, 49
  local, 49, 51, 56, 66
  measure of relative, 68
  models, types 1 and 2, 50
  sign, 53, 66
  species addition, 49
    deletion, 49
stable cycles on hypercube, 83
  subsystems, linked, 54
star, **35**
starting string, 178
stellate cells, 1, 104
stoichiometry, 80
Strahler ordered data, 122
  ordering, 109, 110, 118, 122, 158, 168
    disadvantages of, 158
stress similarity, 163
strictly trophic web, 21
string sequence, 175
subtend, **112**
surface filling tree, 3, 149
switching functions, 80
switching networks, clocked, 89
symmetrical bifurcation, 156
symmetrical branching model, 106, 140,
    165, 211

terminal growth model, 171, 195
Terminal catappa, 149, 195
Thamnidium elegans, 130
thermodynamics, 24, 152
topological units, 170
transitions in network of states, 4, 16, 81
tree, **5**
  binary, 5
  rooted, **5**
  paracladial, 175, 191
  spanning, **43**
trophic impurity, 23, 29
  level, 21
  partition, **44**
  position, fractional, 24, 28
  structure, 20
truth table, 81
turbulent flow, 154

valency, **44**
veins, data in Strahler ordering, 128
ventilation, 152
vertex, **4**
volume filling tree, 3, 132, 140
volume of the lung, 155

Walker–Ashby networks, 92, 98
Weibel ordering, 107, 113, 187
  disadvantages of, 107
weighted digraph, **5**